CLIMATE SECURITY

CLIMATE SECURITY

Prof. Avinash Tiwari

Delve
PUBLISHING

www.delvepublishing.com

Climate Security

Prof. Avinash Tiwari

Delve Publishing
4164 Lakeshore Road
Burlington, ON L7L 1A4
Canada
www.delvepublishing.com

Email: orders@arclereducation.com

ISBN: 978-1-77956-385-9 (Hardcover)

ABOUT THE AUTHOR

Prof. Avinash Tiwari, Vice-Chancellor of Jiwaji University, Gwalior, brings a wealth of academic and administrative experience spanning over 33 years. Prof. Tiwari has held pivotal roles in higher education, including as Professor, Associate Professor, and Assistant Professor. His academic qualifications include NET and a Ph.D., which have underpinned his distinguished career dedicated to teaching, research, and institutional administration. With over 31 years of administrative leadership, he has served as the Head of Department, Examination Controller, Dean of Faculty, and member of UGC and NAAC national bodies. Since his appointment as Vice-Chancellor in December 2021, Prof. Tiwari has steered Jiwaji University to unprecedented achievements, including securing the NAAC A++ Grade, UGC Category I recognition, and a □100 crore grant under the PM-USHA program. His efforts have significantly elevated the university's academic and infrastructural standing while fostering innovation through MOUs with 35 national and international organizations. Under his visionary leadership, Prof. Tiwari has implemented numerous transformative initiatives. These include the establishment of a Medical College and Research Center, the promotion of e-governance, and the enhancement of sustainability efforts within the campus. A staunch advocate for educational accessibility, he has championed initiatives such as the "Bandi Se Bandhu" scheme for educating prisoners and programs to empower tribal communities. Additionally, his specialization in Food Technology, Agro Ecology, and Tourism Management, complemented by his international academic engagements, reflects his commitment to advancing global education and research. Recognized with honors like "Gwalior Ratan" and leadership awards, Prof. Tiwari exemplifies excellence and dedication in the field of higher education.

TABLE OF CONTENTS

LIST OF FIGURES

ABSTRACT

Climate security is an increasingly pressing issue in the modern world. This book explores the intersection of climate change and national security, examining the potential threats posed by environmental changes. It delves into how climate change can exacerbate existing security challenges, such as resource scarcity, health security, and conflict. The book also discusses strategies for addressing these threats, including mitigation and adaptation measures. Overall, it provides a comprehensive overview of the complex relationship between climate change and security, offering insights into how policymakers can effectively respond to this critical issue.

PREFACE

Climate security is a critical concern that extends far beyond the realm of the environment. It has profound implications for global security and affects every facet of our existence. The effects of climate change are extensive and significant, encompassing areas such as food and water security, political stability, and economic development. This book aims to illuminate the interconnectedness of climate change and security, emphasizing the importance of taking a comprehensive approach to addressing these issues.

This book holds significant value for the general public due to its ability to offer a comprehensive and easily understandable analysis of climate security concerns. This book strives to provide readers with a clear understanding of the impact of climate change on security and motivate them to take necessary measures to tackle this issue. It achieves this by simplifying complex concepts and presenting them in a straightforward manner. For anyone interested in gaining a deeper understanding of this urgent matter, this book provides valuable insights that can broaden your knowledge. Whether you're a concerned citizen, a policymaker, or a student seeking to expand your understanding, the information within these pages is sure to be enlightening.

This book offers valuable insights that professionals from various fields can benefit from. It is crucial for policymakers to have a deep understanding of the implications of climate change on security to develop effective strategies to mitigate its impacts. Understanding the security risks associated with climate change is of utmost importance for military and security professionals, as it directly impacts national and global security. Exploring the complexities of climate security can yield valuable insights and innovations for researchers and academics, which in turn can inform future research and policy decisions.

Higher education institutions are crucial in furthering our knowledge of climate security. Through the establishment of a platform for research and discussion, universities, and research institutions play a crucial role in advancing progress in this field. Their efforts contribute to equipping the future leaders with the necessary knowledge and skills to effectively tackle this critical issue. This book seeks to make a valuable contribution to the ongoing dialogue by offering a thorough examination of climate security issues and emphasizing the significance of interdisciplinary collaboration in tackling these challenges.

The material for this book was collected from a diverse array of sources, encompassing academic research, policy papers, and expert interviews. This book seeks to offer a thorough and intricate comprehension of climate security issues by incorporating a wide range of perspectives. This book presents insights that are grounded in the most recent research and analysis in the field. It aims to provide readers with current information and valuable insights to enhance their understanding of this intricate issue.

–Author

CHAPTER 1

INTRODUCTION TO CLIMATE SECURITY

Contents

INTRODUCTION

Climate change is undeniable, catastrophic, and inevitable; the enormous destabilization that may result from its inevitable and uncertain ramifications is yet unknown. Therefore, one of the most important U.S. strategic goals this century must be to stem the tide of climate change and adapt to its far-reaching security implications. We need to think outside the box when it comes to using the military as a tool of national power to reduce the impact of climate change, help states and people who are particularly at risk adapt to the changing environment, and ensure long-term stability in areas where these impacts will be most severe (McDonald, 2013).

The United States needs to address climate change as a security threat, and the time is now. The National Security Strategy, National Defense Strategy, Guiding for the Employment of the Force, and the Quadrennial Defense Review should all deftly incorporate strategies and methods for adapting to and mitigating climate change, as well as its security implications. After 2050 and beyond, when climate change's impacts will have multiplied and intensified, the correct actions must be taken due to the number and kind of destabilizing effects. These new initiatives should be based on the policy of sustainable security, which is a broader definition of national security that places a premium on development and diplomacy. The effects of climate change will be felt for a long time and in many ways, with some areas bearing the brunt of the problems more acutely than others. These difficulties are exemplified by Sub-Saharan Africa, which also presents a distinct opportunity for the Western and Eastern countries (the United States and China, respectively) (Brown, Hammill, and McLeman, 2007).

The US, which has started to press China hard on climate change, should look into military cooperation in Africa as a regional response to the issue. In Sub-Saharan Africa, where the impacts of climate change are already noticeable and are only going to get worse, the United States and China could fortify their vital bilateral relationship through military-to-military cooperation, which would be mutually beneficial. Greenhouse gas emissions are highest in China, the world's most rapidly developing power. The top emitter of greenhouse gases per capita is the United States, the world's leading power (McDonald, 2013).

1.1. CLIMATE CHANGE AND GLOBAL WARMING

Climate change is already here, it's here to stay, and it could get even worse. One of the main causes or contributors is the human-caused phenomenon of global warming, which is caused by the buildup and continued presence of "greenhouse gases" in the atmosphere. Transportation, industry, heat, and electric power generation all rely on the combustion of fossil fuels like petroleum, coal, and natural gas. This process produces greenhouse gases, the most common of which is carbon dioxide (CO_2). It is critical to adapt to a changing climate and to reduce atmospheric concentrations of greenhouse gases, especially in regions where these changes are already having or will have a devastating impact (Figure 1.1).

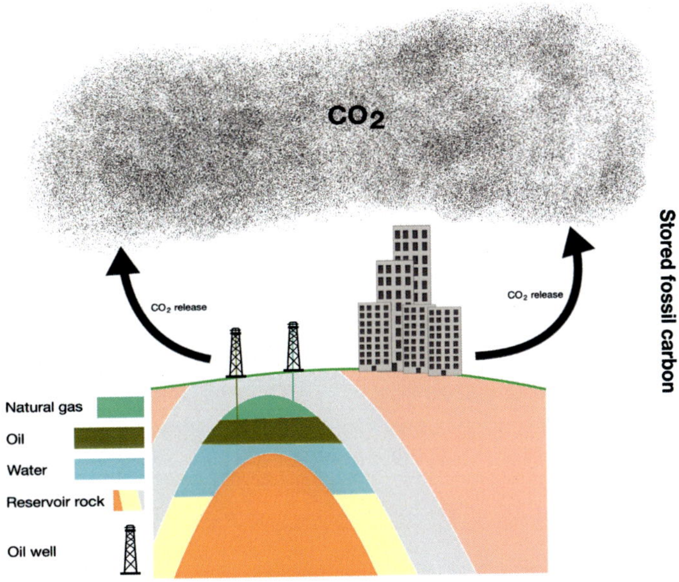

Figure 1.1. *The greenhouse effect.*

Source: Nunes LJR, Meireles CIR, Pinto Gomes CJ, Almeida Ribeiro NMC. Forest Contribution to Climate Change Mitigation: Management Oriented to Carbon Capture and Storage. Climate. *2020; 8(2):21. https://doi.org/10.3390/cli8020021.*

1.1.1. Understanding the Scientific Principles Behind Global Warming.

In 2007, the Fourth Assessment was finally released by the United Nations' Intergovernmental Panel on Climate Change (IPCC). The IPCC report holds significant importance due to its well-balanced and moderate nature. Despite not fully resolving all debates surrounding the topic, climate change is widely acknowledged, from a scientific standpoint, to be a consequence of human-induced global warming. However, there are still uncertainties regarding the specific locations, timing, and extent of its impact. The IPCC report states that there is clear evidence of manmade global warming and that climate change is impacting all continents. It is important to note that the air, oceans, and land are experiencing significant warming, which has undeniable impacts on ecosystems, water resources, human health, industry, settlement, and society as a whole. The effects of climate change will vary across different regions, making them challenging to predict (McDonald, 2013). However, it is expected that climate change will be significantly more pronounced in the coming centuries. Climate change is an ongoing issue that is expected to intensify in the future. Its impacts will persist for a long time, possibly for centuries or even millennia, regardless of the measures taken to address global warming (Boas and Rothe, 2016).

1.1.2. Current and Anticipated Impacts of Climate Change

0In the US and elsewhere, climate change is already having serious consequences. More frequent and more intense heat waves, desertification of broader areas, droughts, and other water shortages that last longer and are more intense, floods that are both more unpredictable and more damaging, wildfires that are both more widespread and more destructive, irreversible rise in sea levels, and an acceleration of biodiversity loss are among the current and predicted consequences. Worldwide, entire island nations may be at risk of flooding due to rising sea levels; in the US alone, this poses a threat to hundreds of millions of coastal residents and billions of dollars in property (Figure 1.2).

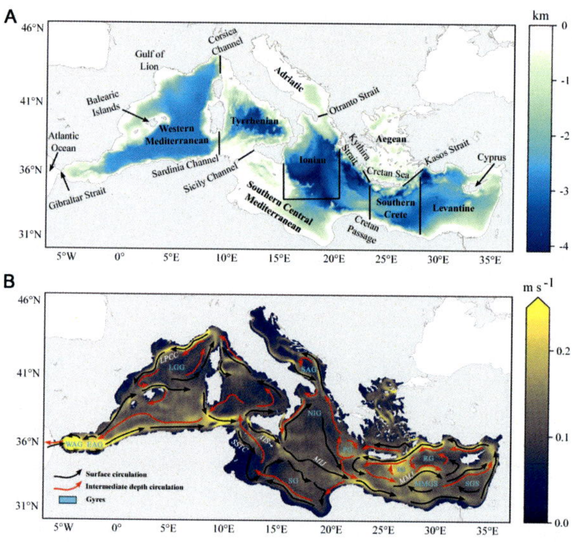

Figure 1.2. *Sea level rise poses 'unthinkable' risks for the planet.*

Source: Meli M, Camargo CML, Olivieri M, Slangen ABA and Romagnoli C (2023) Sea-level trend variability in the Mediterranean during the 1993–2019 period. Front. Mar. Sci. 10:1150488. doi: 10.3389/fmars.2023.1150488.

Over the past 10 years, there has been a tripling of the number of people impacted by natural disasters caused by weather. When combined with the backdrop of fast-growing populations, these events have the potential to cause widespread population displacement, especially due to the unsustainable movement of people from rural areas to cities. Scarce resources may lead to competition. A number of states will collapse, while others will brutally plunder their diminishing resources. At some point, conflict may arise along either path. We are living in an "Energy-Climate Era" where "global warming, global flattening, and global crowding" are coming together (Boas and Rothe, 2016). Estimates of the number of lives lost due to climate change are in the millions, and this has caused widespread alarm.

1.2. CLIMATE CHANGE AND ITS CONSEQUENCES FOR SECURITY

The term "climate security" describes how the climate crisis has threatened stability, especially in already unstable and war-torn regions. As a result of climate change, insecurity in food, water, and livelihoods can become even more severe, leading to a domino effect of:

- Social unrest;
- Displacement; and
- Intensified competition for scarce resources.

A country or region may experience heightened tensions, conflict, and instability as a result of this. Climate change can make it harder to achieve and maintain peace in areas where conflicts are already present by exacerbating or prolonging them. Destroying energy, water, and agricultural assets is one way in which conflict disrupts or impedes climate action. Another way is that it delays or blocks adaptation and mitigation interventions.

1.2.1. How Important is Climate Security?

Understanding the importance of climate security is crucial. By the year 2030, the impact of climate change has the potential to increase the number of people living in poverty by as much as 130 million. This would not only worsen current issues such as lack of access to food and water, but also contribute to socio-economic instability and political unrest. In already delicate environments, these effects can intensify security concerns and exacerbate instability. In addition, it is worth noting that climate-related security risks tend to have a greater impact on women compared to men.

It is important to note that to effectively address climate change, it is crucial that our responses are in line with conflict prevention and peacebuilding actions.

Investing in climate action holds immense importance in fragile and conflict-affected settings. It presents a unique chance to foster cooperation, restore trust, and mend the social fabric if approached with deliberate intentionality. The Sixth Assessment Report from the IPCC acknowledges the potential of climate solutions to foster peace in regions susceptible to climate change-induced conflicts (McDonald, 2013).

In fragile settings, the importance of decentralized renewable energy solutions cannot be overstated. These solutions serve as a lifeline, enabling access to clean water, light, warmth, sustenance, and essential emergency services. They play a crucial role in supporting the overall well-being of these communities. Opting for renewable energy sources and clean technologies during reconstruction efforts not only enhances resilience but also prevents the need for more expensive retrofits in the future (Boas and Rothe, 2016).

1.2.2. Prevalent Climate-Induced Security Risks

1.2.2.1. The Allocation of Scarce Resources

Access to natural resources has been impacted by climate change, leading to increased volatility in availability and distribution. This has significant consequences for communities whose livelihoods and coping strategies rely on these resources. Such circumstances frequently lead to heightened competition for scarce resources and heightened social tensions and conflicts.

On a global scale, the issue of food insecurity has escalated to unprecedented heights as a result of various factors such as conflicts, the COVID-19 pandemic, and the exacerbation of extreme weather events. In the year 2022, a staggering number of approximately 260 million individuals across 58 different countries found themselves grappling with acute food insecurity, while a distressing 35 million people were unfortunately on the brink of starvation.

In countries impacted by conflict and fragility, the consequences are magnified, as individuals are already experiencing humanitarian needs and have limited ability to endure further adversity. In 2022, a staggering 20.9 million people in the Horn of Africa faced extreme food insecurity as a result of the region's longest and most severe drought in recent memory. In Kenya and Somalia alone, nearly 3 million individuals experienced severe food insecurity (Brown, Hammill, and McLeman, 2007).

1.2.2.2. Instances of Insecurity and Polarization

On a global scale, it is evident that the majority of individuals, specifically 6 out of 7, experience a sense of insecurity. This phenomenon can be attributed, at least in part, to the far-reaching consequences of climate change. Climate change has the potential to exacerbate social inequalities and disproportionately impact vulnerable communities, leading to increased social grievances. Disruptions to livelihoods, disputes over resources, and displacement have the potential to exacerbate pre-existing tensions and incite instability (Boas and Rothe, 2016).

In addition, there is a correlation between insecurity and political polarization as well as extremism. According to the Global Peace Index (2022), there has been a significant increase in political instability, violent extremism, strained neighbor relations, and displacement within and across borders over the past 14 years (McDonald, 2013).

1.2.2.3. Mass Displacement

The phenomenon of mass displacement is a complex issue that has far-reaching consequences. According to projections, a staggering 216 million individuals across six different regions may be compelled to relocate within their respective nations by the year 2050 due to the impacts of climate change. The next half-century will see climate relocation being shaped by factors such as water scarcity, low crop productivity, and rising sea levels. This will lead to forced displacement being accelerated, while also intensifying pre-existing tensions or giving rise to new ones.

1.3. CLIMATE SECURITY: A HISTORICAL CONTEXT

Environmental degradation has been a significant concern in terms of security ever since the publication of Lester Brown's influential paper "Redefining National Security" in 1977. In 1989, Jessica Tuchman Matthews published a highly influential article titled "Redefining Security." This article advocated for a more comprehensive understanding of national security, encompassing not only traditional aspects but also resource, environmental, and demographic concerns. Thomas Homer Dixon extensively examined the correlation between natural resources and the occurrence of violent conflict in his articles and books during the 1990s. Jon Barnett's work in 2003, titled "Security and Climate Change," established the foundation for the concept of climate security. In 2007, US defense officials began considering climate change when the Department of Defense's Office of Net Assessment requested a report on the topic. In 2007, the United Kingdom made history by initiating a discussion in the United Nations Security Council that connected energy, climate, and security. In 2007, the Center for a New American Security published "The Age of Consequences: The Foreign Policy and National Security Implications of Global Climate Change," which was later followed by the Council on Foreign Relations' publication titled "Climate Change and National Security: An Agenda for Action." In 2010, the Department of Defense included a statement on climate security in its

Quadrennial Defense Review, emphasizing the importance of climate change and energy in shaping future security. The Department is currently in the process of formulating policies and strategies to effectively address the impact of climate change on its operational environment, missions, and facilities. Another concept to consider is "climate fragility," which pertains to nations that are susceptible to climate change and face interconnected political, social, and economic difficulties (Boas and Rothe, 2016).

1.4. CLIMATE SECURITY IMPLICATIONS

The physical and socioeconomic changes have significant security implications such as:

1. **Violence and Armed Conflict:** The impact of climate change on our planet cannot be underestimated. It will undoubtedly reshape the availability and condition of vital resources like fresh water, arable land, coastal territory, and marine resources. There has been speculation among researchers regarding the potential of these changes to instigate or prolong armed conflict. The connection between the environment and armed conflict is widely recognized. The competition for natural resources, including diamonds, timber, oil, water, and even narcotics, has been a driving force behind violence in various locations such as Kuwait, Colombia, and Afghanistan. Insurgencies in Angola, Sierra Leone, and other regions have been funded, in part, by natural resources. The relationship between climate change and the outbreak of violence may not be as significant as when natural resources are exploited for immediate financial gain. Climate change occurs over an extended period of time, making it improbable for global warming to directly trigger any specific armed conflict. Additionally, the impact of climate change on conflict may not be easily discernible. However, it is important to note that regional climate changes, just like other factors contributing to environmental degradation, have the potential to increase the likelihood of armed conflict (McDonald, 2013).
2. **Occurrences of Catastrophic Events in the Natural World and Situations Involving Widespread Suffering and Urgent Need for Assistance:** A warmer world will increase of natural disasters, consequently leading to a rise in humanitarian crises. Natural disasters pose a significant security threat. It is important to note that a significant number of individuals who experience natural disasters are forced to become refugees or internally displaced persons (IDPs). Refugees and IDPs face a multitude of challenges, including:

a. Detrimental impact of disease;

b. Malnutrition;

c. Loss of income;

d. Risk of personal insecurity; and

e. Crime, violence, and involvement in militarized conflicts.

When countries are unable or unwilling to assist affected populations during natural disasters, it poses a broader security challenge. This not only undermines the perceived legitimacy of local and/or national governments but also fuels popular grievances.

3. **Destability of Failed States:** Governments that aren't strong enough to fix problems like food and water shortages, health crises, population displacement (from rural areas to cities and even across borders), resource, and territorial conflicts, infrastructure damage, and increased poverty (both real and perceived) are likely to lose support from the public. Extremist ideology may flourish in the unforgiving environments brought about by climate change, while transnational crime and the oppression of already vulnerable populations may find these conditions impossible to overcome.

Although major wars have not yet broken out as a result of environmental issues like climate change, there is a growing likelihood of smaller-scale regional conflicts. When outlining the operational environment, the new Army field manual on operations predicts that many developing nations will face even more challenging circumstances due to climate change, which will trigger huge humanitarian crises. The "Joint Operating Environment 2008" report from the United States Joint Forces Command also highlights the significant risks associated with climate change. One estimate puts the number of people forced to flee their homes due to environmental disasters at 200 million by the year 2050, up from 50 million in 2010 (Gemenne et al., 2014).

Anxieties, anger, panic, and distrust will quickly set in as a result of climate change-related consequences such as:

- Mass starvation;
- Pandemic disease;
- Desertification;
- Loss of biodiversity; and
- Rising sea levels;

According to Boas and Rothe (2016) views, maintaining order, delivering humanitarian relief, and creating economic opportunity on the scale that may be necessary will be challenging for governments, whether strong or weak, acting alone or in combination. Potentially disastrous consequences might ripple across nations, areas, and the global system as a whole. Even the most advanced nations are at risk from climate change, but less developed nations will be unable to cope because they lack the resources to adapt. Already, organizations like the Center for a New American Security are working to discover ways to adapt to and lessen the impact of climate change. In July 2008, they held a war game called "Clout and Climate Change: A New Global Agenda for the 21st Century." Many branches of the United States government are worried about the instability that climate change could bring. This includes Congress, the Department of Defense, and the State Department. For instance, the State Department's Bureau of Oceans, International Environmental and Scientific Affairs now houses an Environmental Security Working Group. Along with the Department of State, USAID is becoming more cognizant of climate change issues and is looking for ways to collaborate with the Department of Defense to address the complex interplay between security, stability, and sustainable development through the application of both soft and hard power (Boas and Rothe, 2016).

1.4.1. Most Susceptible States

Some forms of government will be more vulnerable to climate change-related security threats than others around the world. Although the specific regional and local effects of climate change are still hard to foretell, the security risks posed by the phenomenon appear to be most severe for the least developed nations, weak states, and undemocratic states.

1. **Developing Countries:** Perhaps the most vulnerable developing nations will be those that are already struggling economically due to climate change. These nations aren't prepared to deal with climate change because they don't have the resources, infrastructure, or leadership to do so. For instance, they are ill-equipped to deal with humanitarian crises like massive flooding, either to prevent them or to respond to them. Tropical developing nations are bearing the brunt of climate change's harshest impacts, such as increased frequency and intensity of extreme weather, drought, and disease.
2. **Weak Nations:** State instability due to climate change is more likely to occur in failed and failing nations, which are characterized by:

a. Weak government institutions;

b. Lack of border control;

c. Oppressed populations; and

d. Marginal economies.

When it comes to responding to climate change and preventing it from causing a large-scale humanitarian disaster, weak states are practically helpless.

Famine, crop failure, and the subsequent collapse of the Somalian government in the 1990s killed 10s of 1000s of people. The Darfur region of Sudan is already at a high risk of drought, and the country's internal conflict is making matters worse. If these droughts are caused by climate change, but they do fit the pattern that would be expected as a result of global warming.

3. **States That do not Practice Democracy:** Democracy, according to economist Amartya Sen, which has been around for 20 years, does not cause famines because leaders in democracies are answerable to the people who can vote them out of office. Undemocratic governments, on the other hand, failed time and time again in the twentieth century to safeguard populations threatened by natural disasters like drought and floods. There has never been a famine in modern India, but millions of Chinese people perished during Mao's rule. While it can manufacture nuclear weapons, North Korea still can't provide for its people's nutritional requirements. Consequently, the more frequent and severe humanitarian crises caused by climate change will disproportionately affect populations in undemocratic states (McDonald, 2013).

1.5. CHALLENGES IN CLIMATE SECURITY

To effectively tackle the threats to peace and security caused by climate change, it is essential to adopt comprehensive and interconnected strategies. Successful collaboration between the climate and environmental fields and the peace and security ones is crucial for this endeavor.

Coordinating such efforts can pose a significant challenge, given the unique nature of these fields in terms of policy and practice. It is important to note that finding a solution to the complex issue of climate impacts on peace and security is not a straightforward task. The challenges vary significantly between regions such as West Africa, Southeast Asia, and the Pacific Islands. In addition, solely prioritizing emergency response can result in the neglect of crucial long-term investments in adaptation and peacebuilding (Brown, Hammill, and McLeman, 2007). In addition, the financial resources required for this project are currently fragmented and inadequate to fulfill the requirements. Take countries in fragile settings, for instance. These nations bear the brunt of severe climate impacts, yet they receive disproportionately low levels of climate finance. According to a study conducted by UNDP, there is a significant disparity in the amount of multilateral climate finance received by extremely fragile states compared to non-fragile states. The study reveals that between 2014 and 2021, extremely fragile states received a mere US$2.1 per person, whereas non-fragile states received a much higher amount of $161.7 per person (Boas and Rothe, 2016).

1.6. RECOMMENDATIONS FOR CLIMATE SECURITY ISSUES

The strategy employed by the United Nations to address climate change involves facilitating agreements between nations. The aim is to reduce greenhouse gas emissions to stabilize atmospheric concentrations of these gases at a safe level. Additionally, the United Nations seeks to assist vulnerable nations in adapting to the adverse consequences of global warming. Although the goals are indeed commendable, it is evident that the UN system lacks the necessary level of ambition and effectiveness to effectively address the security risks associated with climate change. Below are several new approaches.

1.6.1. Addressing Emissions

Only when yearly net global emissions of greenhouse gases fall to zero will atmospheric concentrations of these gases stabilize, and until then, global warming will persist.

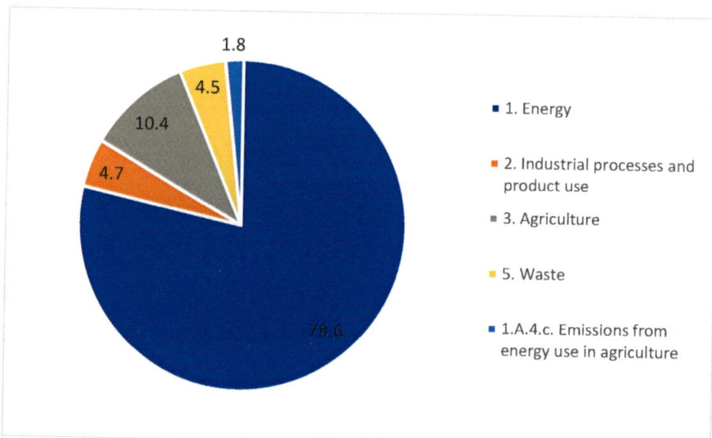

Figure 1.3. *The largest emitters of greenhouse gases.*

Source: Genstwa N, Zmyślona J. Greenhouse Gas Emissions Efficiency in Polish Agriculture. Agriculture. *2024; 14(1):56. https://doi.org/10.3390/agriculture14010056.*

The United Nations' climate-change security strategy should prioritize helping both developed and developing countries reduce their emissions since most countries are still seeing fast increases in emissions. The United

Nations Framework Convention on Climate Change of 1992 and the Kyoto Protocol of 1997 were the primary frameworks within which international efforts to mitigate climate change were implemented. So far, the outcomes of those endeavors have been quite meager. Major industrialized nations failed to fulfill their convention-mandated pledge to reduce emissions to 1990 levels by the year 2000. Not only will the Kyoto Protocol not apply to the biggest national emitters—China and the United States—but it will only cover 25% of worldwide emissions even if it is implemented (Figure 1.3).

Furthermore, by 2012, the Kyoto Protocol was to have achieved a reduction in emissions of less than 3% below 1990 levels in the countries that had participated. Climate change will outstrip most governments' adaptive measures unless new, substantial efforts are made to reduce its impact. Both inside the North-South dialogue and among developed industrialized nations, the Kyoto Protocol's climate architecture has grown more contentious. Politically, environmental issues have also declined since the Earth Summit of 1992, when they were absorbed into the larger sustainable-development agenda. If the Secretary-General wants to see mitigation efforts picked up speed, he or she needs to make climate change a bigger deal and help remove barriers to reducing emissions.

While wealthy countries should spearhead efforts to curb emissions, developing countries may find more cost-effective ways to do it. This adds another layer of complexity. The world will most certainly keep warming until its leaders muster the political will to reduce emissions through a combination of public and private funding (McDonald, 2013).

Because of more pressing security concerns, like Iraq, climate change gets less attention than other, longer-term security concerns. A more official institutional framework is required to consistently address the issue of climate change, but the Secretary-General should make it a more integral part of his personal diplomacy.

The Secretary-General could, for example, push for the establishment of a United Nations High Commissioner for the Environment. The role of the High Commissioner would be to bring attention to the worst and best environmental practices while simultaneously raising awareness about environmental degradation on a global scale, including climate change. By moving the office to Geneva, the United Nations could finally address climate change and other environmental issues, something that the UN Environment Program in Nairobi has failed to do. The stance would represent a middle ground between countries that are in favor of stronger global environmental governance and those that are opposed, including France and Germany.

1.6.2. Adaptation

Adaptation involves the ability to adjust and modify behaviors, physical characteristics, and physiological processes to better suit changing conditions.

Even if the mitigation agenda is successful, atmospheric concentrations of greenhouse gases will continue to climb for some time. These concentrations are at their highest point in 10s of 1000s of years. To combat climate change's negative impacts, a two-pronged approach is required.

To start, the United Nations should put more resources into its disaster and humanitarian relief programs, especially those that are starting to factor in climate change. In addition, the United Nations should launch a fresh initiative to better anticipate, avert, and deal with climate-related catastrophes in nations that are politically unstable or have authoritarian regimes (Gemenne et al., 2014).

1.6.2.1. Fortifying Ongoing Disaster Response

1.6.2.1.1. Refocus Attention From Relief to Prevention

When it comes to responding quickly to new disasters, humanitarian groups have gotten much better. Nevertheless, disaster risk reduction receives a pitiful allocation of funds. Too little is known about how important it is to reduce disaster risks, even in nations with responsive decision-makers. At least 5%–10% of humanitarian relief funds should be set aside for disaster-risk reduction (Boas and Rothe, 2016).

While it is ultimately up to the member states to decide on this specific objective, the Secretary-General should spearhead the effort to establish such a principle. Both the Inter-Agency Secretariat of the International Strategy for Disaster Reduction (UNISDR) and the UN's Inter-Agency Task Force on Disaster Reduction (IATF/DR) are preexisting frameworks that incorporate vulnerability assessments and early-warning systems to varying degrees.

1.6.2.1.2. Incorporate Climate Change and Disaster Preparedness

Security, disaster-prevention, and humanitarian-response efforts within the United Nations system should give greater weight to worries about climate change's potential effects. The United Nations should prioritize climate change in its UN/ISDR efforts, for instance. A project with similar goals was initiated by UN/ISDR in 2003, and so far, it appears to be making good progress. As an example, a new working group on climate adaptation and disaster reduction was established in May 2004 by the IATF/DR. The UN/ISDR Secretariat is also coordinating an expert dialogue among the development, climate, and disaster relief communities. Despite the growing importance of adaptation within the climate change community, organizations working to address this issue run the risk of starting from scratch when it comes to disaster preparedness and response. To circumvent this possible issue, the current network of catastrophe specialists should be better integrated into the IPCC reporting process.

1.6.2.2. A Fresh Approach is Necessary for States in Danger

According to McDonald (2013), the most dangerous security threats, such as large migration, armed conflict, and state collapse, are more likely to happen in undemocratic and weak states; even with improved UN early-warning and disaster-preparedness systems, these systems would not be completely sufficient to anticipate and handle these crises. Tackling these challenges calls for a fresh, multi-pronged approach.

1.6.2.2.1. A Better Vulnerability Index and Early Warning System

The United Nations system requires more accurate methods of identifying which areas and states face the greatest climate-related security risks. The international community has already seen a proliferation of early warning systems to address various threats. Among the many systems used in the humanitarian sphere, one such system is the United Nations' Humanitarian Early Warning System (HEWS), which is used internally to detect nations in pre-crisis states.

An external system that specializes in dealing with complex emergencies and natural disasters, Reliefweb is also supervised by the United Nations' Office for the Coordination of Humanitarian Affairs (OCHA). In order to better serve countries and regions in sub-Saharan Africa, OCHA has established an Integrated Regional Information Network (IRIN). When it comes to agriculture, there are two systems that can be relied on: GIEWS, which is run by FAO, and Famine Early Warning System, which is run by USAID.

The United Nations Development Programme and the United Nations Environment Programme/GRID have created a Disaster Risk Index (DRI) for weather-related warning systems. There is a lot of good in this work. However, political indicators of vulnerability, such as (a) the oppressive character of political regimes and (b) other governance factors, need to be more thoroughly considered by the developing early-warning systems within the disaster-reduction community.

When it comes to standardizing risk assessments in a format that policymakers can use, coordination among the aforementioned vulnerability indexes and other early warning systems is essential. One such initiative is the European Union's Global Monitoring for Environment and Security (GMES). Finally, the IPCC needs to conduct a thorough analysis of the climate-security nexus. A gathering of international experts in climate change, disaster preparedness, and security should be called by the IPCC as an initial step (Gemenne et al., 2014).

1.6.2.2.2. Conflict Avoidance Diplomacy

After the United Nations has determined which nations are most at danger, it needs to make preparations to deal with the aftermath of climate change. It is unclear what level of systematic forward planning the United Nations is currently engaged in. The first thing to do is make sure that any backup plans for those countries' shelter, food, medicine, and police are specific to their own situations. Also, local UN staff (or special envoys, if needed) should establish separate lines of communication with decision-makers in high-risk nations to promote risk-reduction measures and discuss them. The United Nations High Commissioner for Refugees, the International Red Cross, and other non-governmental organizations (NGOs) that provide aid should also receive updates from UN officials on disaster prevention efforts (McDonald, 2013).

1.6.2.2.3. Legitimacy and Force in Post-Conflict Engagement

There will be times when diplomatic readiness isn't enough to avert a humanitarian disaster, and the world will have to consider the possibility of resorting to force to avert mass famine or destabilizing migrations. Therefore, discussions regarding a permanent multilateral peacekeeping or humanitarian intervention force should take climate change's security risks into account. Similar to how it is addressing weapons of mass destruction and terrorism, the international community must reevaluate its institutional arrangements and norms regarding the use of force in the aftermath of disasters. In addition to bolstering its post-conflict reconstruction programs to encompass climate-induced disasters, the United Nations should be facilitating this conversation (Boas and Rothe, 2016).

1.7. SUSTAINABLE CLIMATE SECURITY

It is crucial that we address the challenges posed by climate change and develop strategies to ensure the long-term security of our planet. Climate change is a global issue that requires a comprehensive and coordinated approach. By implementing sustainable practices and adapting to the changing climate, we can safeguard our future and create a more secure world for generations to come.

The three pillars of sustainable security—human, collective, and national—must be upheld simultaneously. In doing so, it broadens the conventional, inward-looking idea of state self-preservation to include transnational dangers that are pernicious to an increasingly globalized international system. Sustainable security combines development, diplomacy, and defense in a proactive and preventive manner. It promotes leadership and moral persuasion to address the causes of global violence and instability through development-based solutions rather than relying solely on coercion and force. The Center for American Progress, a prominent proponent of sustainable security, claims that climate change is "a threat multiplier in some of the most volatile regions of the world," citing a 2007 study from the Center for Naval Analyses.

In terms of U.S. combatant commanders' involvement in sustainable development, long-term planning is fundamental to sustainable security. In other words, there is a need for the US military to be more deeply and broadly involved in development assistance.

Along with the Department of State, USAID, and other U.S. agencies, the U.S. military must have the critical strategic capabilities to meet basic human needs and promoting good governance. Lessons from the Horn of Africa, Iraq, and Afghanistan make this clear, as does what might be revealed by developments in Pakistan's ungoverned tribal areas (Brown, Hammill, and McLeman, 2007).

To adapt to and mitigate the effects of climate change, a combination of sustainable development and sustainable security will be required, linked by grand strategy and military strategy. The human, financial, and ecological costs of climate change will be felt heavily by the political, economic, and military institutions of every society, exacerbating already precarious social conditions. Africa might serve as the best example.

Climate change is a pressing and compelling context for long-term safety, particularly in sub-Saharan Africa. Unfortunately, given the size of the issue and the wide range of national interests at stake, the United States cannot and should not address it alone.

The United States ought to take China's interests into account in addition to the regional and national interests of African states.

Furthermore, China's significant interests in Africa offer the US a rare and promising chance to advance both its security interests in Africa and with China at the same time. It's also possible that China is becoming more influential in Africa at the expense of the United States. In certain African nations, China is already regarded as a significant global force. Not because they love China per se, but rather because China's continued presence in the region acts as a check on the unpleasant and aggressive policies of the United States, which is appreciated by some Africans (Brown, Hammill, and McLeman, 2007).

CHAPTER 2

THE SCIENCE OF CLIMATE CHANGE

Contents

2.1. UNDERSTANDING THE FACTORS BEHIND CLIMATE CHANGE

The fundamental cause of climate change lies in the greenhouse effect, which refers to the process by which specific gases in the atmosphere retain heat that would otherwise escape into space from the Earth's surface. One must acknowledge the crucial role played by the greenhouse effect in sustaining life on our planet. Without it, Earth would be a frigid and inhospitable place. However, during the mid- to late 19th century, human activity started to significantly contribute to the intensification of the greenhouse effect. Currently, the planet is experiencing unprecedented warmth, surpassing any previous period in human history, and this trend is continuing to intensify. Global warming has resulted in significant changes to natural cycles and weather patterns. These changes have led to various impacts such as:

- Extreme heat;
- Prolonged drought;
- Heightened flooding;
- Intensified storms; and
- Elevated sea levels.

Collectively, these unfortunate and occasionally fatal consequences are commonly referred to as climate change.

Exploring and analyzing the human causes of climate change is not intended to shame individuals or make them feel guilty for their choices. The key is to clearly identify the problem to find viable solutions. It is imperative that we address the origins of the matter, despite the potential challenges and discomfort that may arise. Throughout history, human civilization has witnessed remarkable advancements in productivity, which unfortunately have contributed to the current issue of global warming. Through the utilization of our innovative capabilities and a collective sense of responsibility, we have the potential to mitigate global warming, combat climate change, and pave the way for a future that is fair, inclusive, and environmentally sustainable (Basics of Climate Change, US EPA, 2024).

2.1.1. Causes That Occur in Nature

Natural events are responsible for a portion of the observed climate change. Volcanic eruptions, variations in solar radiation, tectonic movements, and even minute variations in Earth's orbit have all been observed to impact the planet's warming and cooling patterns throughout Earth's history. Recent global warming, especially that which has occurred since the beginning of the industrial revolution, is occurring at an unprecedented rate. Although these natural causes are still present, NASA states that they either have too little of an impact or happen too slowly to account for the remarkably fast warming that has been observed in the last several decades. Furthermore, contrary to what some in the fossil fuel sector and conservative think tanks would have us believe, the records disprove the idea that natural causes are the primary drivers of climate change (Climate Change Science, US EPA, 2024).

2.1.2. Climate Change Caused by Humans

According to Knight (2023), everything that is happening on Earth right now is mostly caused by humans. Over the last 150 years, atmospheric concentrations of greenhouse gases, including carbon dioxide, have risen dramatically due to the unregulated combustion of fossil fuels. Carbon sinks are natural resources that store carbon dioxide and prevent its release into the atmosphere. However, wetlands, forests, and other carbon sinks have been severely damaged or destroyed due to development and logging. We are currently experiencing the highest atmospheric concentrations of greenhouse gases in the last 800,000 years, including carbon dioxide, methane, and nitrous oxide. Hydrochlorofluorocarbons (HFCs) are one example of a greenhouse gas that is completely absent in the natural environment. The present age of deadly and increasingly common weather extremes is a direct result of the continuous pumping of these gases into the atmosphere, which helped raise the earth's average temperature by approximately 1.9°F during the 20th century. Also, while everyone feels the effects of climate change, it's important to remember that the impact isn't uniform: Although these communities contribute significantly less to global warming, people of color and those from economically or politically marginalized backgrounds endure a disproportionately heavy load (Basics of Climate Change, US EPA, 2024). The main factors that contribute to climate change included.

2.1.2.1. Transportation

Greenhouse gas emissions from transportation—including cars, trucks, ships, and planes—are a big problem on a global scale. Combustion engines, which use fuel derived from petroleum, emit large quantities of carbon dioxide when they burn. About 50%–60% of all emissions come from passenger vehicles; annually, the average passenger vehicle releases around 4.6 metric tons of carbon dioxide into the atmosphere. In terms of road pollution, trucks are in the vanguard. While trucks only make up 4% of the vehicles in the US, they are responsible for 20%–25% of all transportation-related greenhouse gas emissions due to their constant operation and heavy reliance on diesel fuel. We can reduce these figures, but only through massive investments in public transportation and the proliferation of zero-emission vehicles (Figure 2.1).

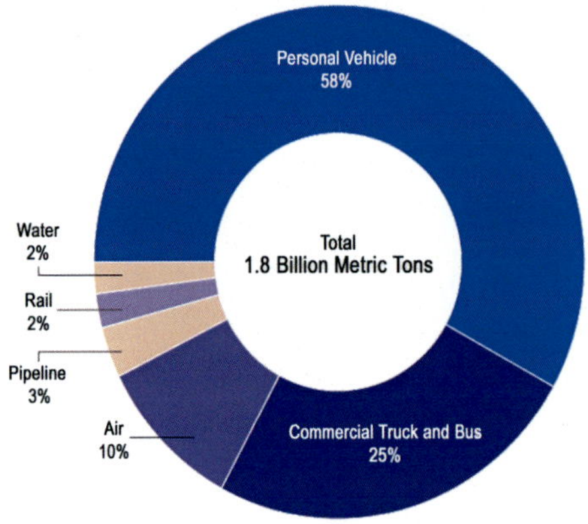

Figure 2.1. *Motor vehicles—personal vehicles and commercial trucks and buses—accounted for 83% of emissions in the transportation sector in 2019.*

Source: Congressional Budget Office, 2022.

2.1.2.2. Industrial and Manufacturing Sector

In 2020, greenhouse gas emissions in the United States were entirely attributable to the facilities and factories that manufacture our goods. A small number of carbon-intensive products—basic chemicals, iron, and steel, cement, and concrete, aluminum, glass, and paper—constitute the majority of industrial emissions. The components of our infrastructure and the wide variety of consumer goods require enormous quantities of energy to be produced. Another common source of these gases and other dangerous air pollutants is the leakage of older facilities that require efficiency upgrades (Knight, 2023).

Enhanced technological efficiency and stricter enforcement of pollution laws could help the industrial sector lessen its impact on the environment. We can also reduce the need to produce so many things by changing our perspective on consumption, recycling, and reusing—especially with regard to plastics.

Furthermore, policy mandates must utilize the purchasing power of the government to promote cleaner alternatives, and that state and federal agencies acquire materials that are sustainably produced because large-scale infrastructure projects are dependent on industries such as cement manufacturing, which contributes to 7% of the world's greenhouse gas emissions each year. To further reduce pollution in this energy-hungry industry, we must move quickly away from fossil fuels and toward renewable energy sources (Climate Change Science, US EPA, 2024).

2.1.2.3. Gas and Oil Exploration

The extraction of oil and gas, as well as their subsequent combustion for fuel, results in emissions at every point in the production and consumption chain. Methane seeps out of the earth at every stage of the fossil fuel industry's production process, including drilling, fracking, transportation, and refining. In the first two decades after being released into the atmosphere, methane is a far more effective heat trap than carbon dioxide, despite the fact that it isn't a greenhouse gas as commonly found. Methane leaks out of even non-running wells, which are called "orphaned" wells. There are over 3 million of these old, inoperable wells all over the nation, and in 2018, they released over 280,000 metric tons of methane.

2.1.2.4. Power Generation Sector

Fossil fuels such as coal and natural gas account for nearly 60% of the US electricity consumption. Roughly 25% of the United States' emissions of greenhouse gases, including carbon dioxide, methane, and nitrous oxide, come from the power sector, which has historically invested in these polluting energy sources. But that past is changing dramatically: Wind and solar power are becoming more popular among utilities as they are becoming more cost-effective and easier to develop. Pollutant levels are falling in tandem with the increasing share of clean, renewable energy. Things are heading in the right way, but not quickly enough. Scientists warn that if we want to avoid the worst climate change impacts, the world's average temperature must not rise more than 1.5°C. To achieve this goal, we must seize every opportunity to rapidly transition the power sector away from fossil fuels and toward renewable sources of energy (Basics of Climate Change, US EPA, 2024).

2.1.2.5. The Field of Agriculture

In 2020, 11% of the United States' greenhouse gas emissions came from agriculture, a clear indication of how drastically modern, industrialized farming has changed the delicate but crucial connection between soil and climate. The significant quantities of the powerful gases nitrous oxide and methane, which are excellent heat traps, that are released by this industry are particularly well-known. Agriculture is responsible for approximately 75% of the nitrous oxide in the atmosphere due to the extensive use of chemical fertilizers and specific crop-management practices that put an emphasis on high yields at the expense of soil health. Methane is a gas that is released into the atmosphere as a result of the digestion of ruminants like cattle and other large-scale industrialized livestock producers. Indigenous farmers in particular have been sustainably tending their land, and their stories serve as a reminder that there is more than one solution to the world's food crisis. Reduce emissions from this industry, increase soil carbon sequestration, and improve food quality by embracing regenerative agriculture's guiding principles and practices (Climate Change Science, US EPA, 2024).

2.1.2.6. Forest Loss

Clearcutting forests and degrading wetlands is another way humans are increasing atmospheric concentrations of greenhouse gases. By retaining it at or below ground, vegetation, and soil serve as carbon sinks. Deforestation, mining, and logging all contribute to the loss of plant and animal life and the subsequent release of carbon into the atmosphere.

Every year, clearcutting in Canada's boreal forest releases over 25 million metric tons of carbon dioxide into the atmosphere—equivalent to the emissions from 5.5 million vehicles. To counteract this trend and replenish Earth's carbon sinks, we need sustainable practices-focused government policies and changes in consumer behavior.

2.1.2.7. Building Structures

Unsurprisingly, given how much time we spend inside of them, our buildings—both residential and commercial—emit a lot of greenhouse gases. Heating, cooling, cooking, running appliances, and maintaining other building-wide systems accounted for 13% of U.S. emissions overall in 2020.

And even worse, some 30% of the energy used in U.S. buildings goes to waste, on average. Every day, great strides are being made in energy efficiency, allowing us to achieve the same (or even better) results with less energy expended. By requiring all new buildings to employ the highest efficiency standards—and by retrofitting existing buildings with the most up-to-date technologies—we'll reduce emissions in this sector while simultaneously making it easier and cheaper for people in all communities to heat, cool, and power their homes: a top goal of the environmental justice movement (Leary, et al., 2001).

2.1.2.8. Our Lifestyle

0Our individual carbon footprints are the result of the choices we make on a daily basis, such as:

- The things we buy;
- The amount of power we use;
- The mode of transportation we use; and
- The food we eat (food waste accounts for 4% of total U.S. greenhouse gas emissions).

The total human carbon footprint is the sum of all these individual footprints. Recognizing the unequal distribution of climate change's causes and effects is the first step toward reducing it.

Those responsible for the majority of the world's greenhouse gas emissions should cut back without harming the least responsible. In the end, our wants, needs, and priorities determine the major climate-altering decisions made by utilities, businesses, and governments. We need to reevaluate those demands, increase their intensity, and rearrange their priorities if we want to win the battle against climate change. We need to shift from corporate-driven, short-termism to community-focused, long-term planning that ensures everyone's health and safety.

We need to quickly shift from using catchphrases to taking collective action in response to climate change. This action should be driven by a sense of righteous anger, but also by a firm belief in science and our capacity to make a positive difference. If our actions have led us to this precarious juncture in human history, then shedding familiar habits will be key to discovering a solution (Basics of Climate Change, US EPA, 2024).

2.2. GREENHOUSE GAS EMISSIONS

Atmospheric gases that act as heat traps are called greenhouse gases (GHGs). At midday, the sun's rays penetrate the atmosphere and warm the surface of the Earth. As the sun goes down, the Earth's surface cools and radiates heat back into space. The atmospheric greenhouse effect, however, keeps some of the heat from escaping.

The average temperature of the Earth remains at 14°C (57°F) because of that. The gases are named "greenhouse gases" because they mimic the effect of a greenhouse's glass walls. If the greenhouse effect did not exist, the Earth's temperature would plummet to -18°C (-0.4°F), which is inhospitable to life (Knight, 2023).

However, the natural greenhouse effect is being altered as a result of the substantial increase in greenhouse gas emissions caused by human activities. When it comes to climate change and global warming, scientists are in agreement: greenhouse gases are to blame.

More and more greenhouse gases have been released into the atmosphere by humans since the Industrial Revolution. Because of the domino effect of climate change, that sum has risen sharply during the last hundred years. Within the last 30 years, the world's average temperature has accelerated to its highest point since records began.

2.2.1. The Primary Greenhouse Gases

2.2.1.1. Carbon Dioxide (CO_2)

About three-quarters of all emissions come from CO2, making it the most important greenhouse gas. Its atmospheric persistence can be 1000s of years.

Some examples of natural processes that release CO_2 into the atmosphere are:

- Volcanic eruptions;
- Plant respiration; and
- The breathing of animals and humans.

However, since the start of the Industrial Revolution in the 1800s, human activities such as burning fossil fuels and massive deforestation have caused the atmospheric CO_2 concentration to rise by 50%. CO2 is the most important gas in changing the climate because it is abundant (Knight, 2023).

2.2.1.2. Methane

Decomposition is an inherent process that produces methane. Unfortunately, the natural balance has once again been disrupted by human activity. Cattle raising, landfills, rice paddies, and conventional oil and gas extraction all contribute significantly to global warming. Even though methane molecules only remain in the atmosphere for around 12 years compared to carbon dioxide molecules, their potency increases by a factor of at least 84 over the course of 20 years. It is responsible for around 16% of the world's total emissions of greenhouse gases (Climate Change Science, US EPA, 2024).

2.2.1.3. Nitrous Oxide (N_2O)

Nitrous oxide accounts for only around 6% of all greenhouse gas emissions, but the IPCC reports that its lifetime in the atmosphere is more than a century and that it is 264 times more potent than carbon dioxide over 20 years. The primary causes of nitrous oxide emissions, aside from the combustion of fuel, are activities related to agriculture and cattle, which include the use of fertilizer, manure, and the combustion of agricultural residues.

2.2.1.4. Industrial Gases

Fluorinated gases, which remain in the atmosphere for 100s–1000s of years, can trap heat 1000s of times more effectively than carbon dioxide. Examples of these gases are hydrofluorocarbons, perfluorocarbons, chlorofluorocarbons, and nitrogen trifluoride (NF3). They make up around 2% of all emissions and are produced as by-products of manufacturing, solvents, and refrigerants (Knight, 2023).

2.2.1.5. Water Vapor

The most common greenhouse gas in the world is actually water vapor, but because it is not directly released by human activity and its effects are poorly understood, it is not monitored in the same manner as other greenhouse gases (Figure 2.2).

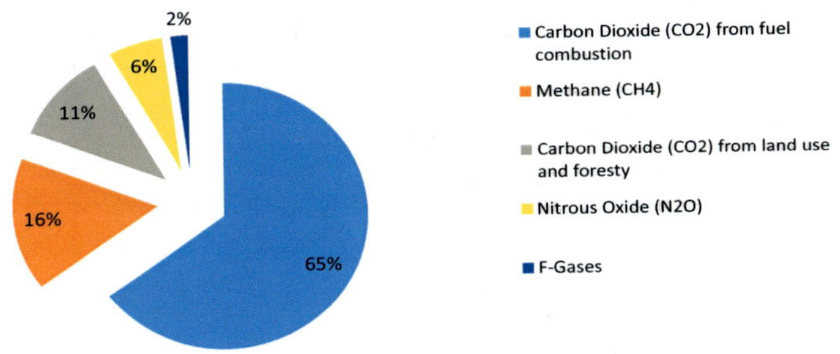

Figure 2.2. *Five major greenhouse gases.*

Source: Cifci E, Oliver ME. Reassessing the Links between GHG Emissions, Economic Growth, and the UNFCCC: A Difference-in-Differences Approach. Sustainability. *2018; 10(2):334. https://doi.org/10.3390/su10020334.*

2.2.2. The Impact of Greenhouse Gases

The devastating impacts of greenhouse gases on ecosystems and human health are far-reaching.

They amplify the effects of smog and air pollution, which in turn cause respiratory diseases, and they accelerate climate change by retaining heat. Additional consequences of greenhouse gas-induced climate change include more frequent and severe weather events, problems with food supplies, and an increase in wildfires. Some species will go extinct, while others will move or expand, and the usual weather patterns we've come to expect will shift.

2.2.3. Strategies for Lowering Emissions of Greenhouse Gases

To save the planet from the worst of climate change, we need to find ways to power our manufacturing, agriculture, transportation, and electricity generation operations in a way that doesn't rely on fossil fuels. The 2015 Paris Climate Agreement was a global acknowledgment of this fact. 20 nations account for over three-quarters of global greenhouse gas emissions; the top three emitters are India, China, and the US. The largest emitters will feel the effects of the changes the most (Basics of Climate Change, US EPA, 2024).

Greenhouse gas emission reduction technologies are largely already in existence. Some of these measures include:

- Increasing energy efficiency;
- Shifting away from fossil fuels to renewables; and
- Pricing carbon emissions to discourage their production.

To keep global warming below 1.5°C, the world has a mere 20% of its "carbon budget" (2.8 trillion metric tons) left. Eliminating fossil fuels won't be enough to reverse the current trends. Actually, implementing strategies to capture CO_2 from the air is an essential component of any plan to limit the world's average warming to 1.5°C or 2°C, the two targets set by the IPCC. Some examples include:

- Reforestation efforts;
- The preservation of existing grasslands and forests; and
- The capture of carbon dioxide gas from industrial and power sources.

2.3. CLIMATE MODELS

The possible effects of climate change, such as shifts in temperature, precipitation, and sea level, can be better understood with the use of models. As per the NOAA Climate Prediction Center, a climate model is a mathematical model that is used to quantitatively describe, simulate, and analyze the interactions between the atmosphere and the underlying surface, such as the ocean, land, and ice. Multiple models of the Earth's surface, atmosphere, and oceans come together in a Global Climate Model (GCM). The Earth is divided into numerous layers and 1000s of gridded spaces in three dimensions by GCMs. These simulations are very good at capturing the present and past weather. Extreme weather events, sea ice dynamics, and warming trends over time can all be reproduced by GCMs. According to the scenarios, the climate models predict potential changes in the weather in the future. These models are iterated with different future scenarios that include different population levels and expected emissions of greenhouse gases like CO2 (Leary, et al., 2001).

When it comes to greenhouse gas emissions, each GCM is unique and sensitive in its own way. Researchers rely on this range, when aggregated, to get a feel for the degree of uncertainty surrounding potential future events within a given scenario and time frame. The complement of projections and this range can be captured by using ensembles of several global climate model simulations. Phase 3 of the Climate Model Intercomparison Project (CMIP3), an ensemble set of global climate model simulations, was utilized for the Fourth Assessment Report (AR4) by the Intergovernmental Panel on Climate Change (IPCC). The third report of the United States National Climate Assessment

(NCA) also made use of CMIP3. For their 5th Assessment Report, the IPCC utilized an updated ensemble, the Coupled Model Intercomparison Project half 5 (CMIP5). Near-term (up to 2035) and far-term (out to 2100 and beyond) climate change projections will be made available by CMIP5, which promotes a standard set of model simulations. Updated CMIP5 models employ a new set of scenarios for greenhouse gas emissions and generate higher-resolution projections than CMIP3.

As a whole, the CMIP3 and CMIP5 forecasts for the United States are quite similar. There are, however, key distinctions concerning particular processes (precipitation) and locations (e.g., California, the southern United States). The data will be the most accessible and easy to use for health departments because a variety of climate products (projections and outlooks) have already been created using CMIP3. Therefore, current materials using CMIP3 should be utilized if capacity and/or capability are limiting factors. The most current set of widely used climate models is CMIP5, so state and local health departments should utilize it whenever possible if modeling capacity exits. This is since CMIP3 studies will soon become outdated. A supplementary set of downscaled (high geographic resolution) climate models may be necessary to produce more accurate climate predictions for certain public health outcomes. In such instances, health departments might have to rely on statistically downscaled predictions from CMIP3 or another climate model, or use a regional climate model. Departments of health that use CMIP5 must possess the necessary technical skills to manipulate, comprehend, and apply the results of climate models. Universities, state climatologists, NOAA's RISA programs, and other relevant organizations may need to work together on this (Basics of Climate Change, US EPA, 2024).

2.4. CLIMATE FEEDBACK AND SENSITIVITY MECHANISMS

Climate feedback mechanisms involve complex interactions within the Earth's system that can either amplify or dampen the effects of climate change. These mechanisms play a crucial role in shaping the long-term climate patterns we observe. By understanding how these feedbacks operate, scientists can gain valuable insights into the potential future impacts of climate change.

In regards to the Earth's climate system and specifically its reaction to extra-terrestrial radiation forcing, several feedback mechanisms are crucial. The emission of longwave radiation into space from Earth's surface is the most basic feedback mechanism. According to the Stefan-Boltzmann law, this radiative loss grows as surface temperatures do, thus it acts as a stabilizing factor (a negative feedback) for the air temperature close to the surface.

The amount of surface warming that results from each additional watt per square meter of radiative forcing is known as climate sensitivity. Another way to look at it is the amount of heat that would be released into the atmosphere if CO_2 levels were to double along with an additional 4 watts per square meter of radiative forcing. For every extra watt per square meter of radiative forcing, the climate sensitivity would be around 0.25°C (0.45°F) if there were no additional feedbacks (Knight, 2023). On the other hand, a 1°C (1.8°F) rise in air temperature would be the consequence of an extra 4 watts per square meter of radiative forcing, which would be the equivalent of doubling the atmospheric CO_2 concentration from 280 ppm at the beginning of the industrial age to 560 ppm. Nevertheless, there are other feedbacks that have a destabilizing effect instead of a stabilizing one, and these feedbacks typically raise the climate's sensitivity to an additional 0.5°C –1.0°C (0.9°F –1.8°F) for every additional watt per square meter of radiative forcing (Leary, et al., 2001).

2.4.1. Cloud Cover Feedbacks

There is a widely accepted belief that when the Earth's surface warms and the amount of water vapor in the atmosphere rises, there is a corresponding increase in global cloud cover. Nevertheless, the impact on near-surface air temperatures is quite intricate.

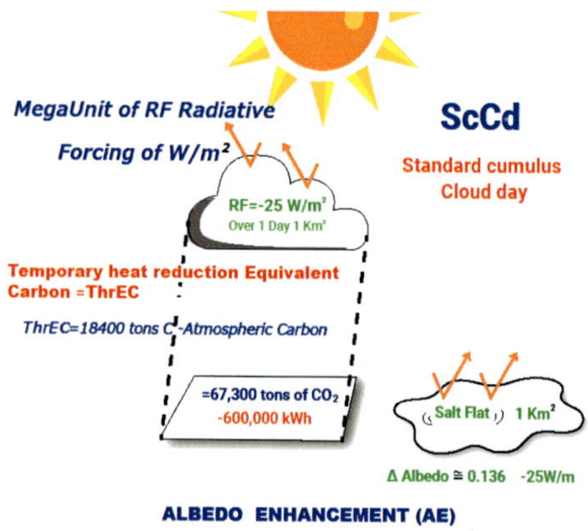

Figure 2.3. *Schematic showing how cloud types affect Earth's heat budget.*

Source: Lightburn KD. Can a Symbolic Mega-Unit of Radiative Forcing (RF) Improve Understanding and Assessment of Global Warming and of Mitigation Methods Using Albedo Enhancement from Algae, Cloud, and Land (AEfACL)? Climate. *2023; 11(3):62. https://doi.org/10.3390/cli11030062.*

When it comes to low clouds, like marine stratus clouds, the most significant radiative characteristic of the cloud is its albedo. In this scenario, an increase in low cloud cover has a similar effect to an increase in surface ice cover. It causes more solar radiation to be reflected, resulting in a cooling of the Earth's surface. However, it is important to note that high clouds, like the towering cumulus clouds that reach the boundary between the troposphere and stratosphere, have a distinct effect on the surface radiation balance. Cumulus clouds have significantly higher altitudes and lower temperatures at their tops compared to their undersides. The tops of cumulus clouds release less longwave radiation into space compared to the warmer cloud bottoms that emit radiation towards the surface. High cumulus clouds contribute to increased surface warming as a result of their formation (Figure 2.3).

According to Knight (2023), the overall impact of clouds on increasing surface temperatures remains somewhat uncertain. This scenario involves a competition between the effects of high and low clouds, making it challenging to establish a clear balance. However, it is widely believed that clouds generally contribute to positive feedback, resulting in further warming.

2.4.2. Carbon Cycle Feedbacks

An essential aspect to consider is the global carbon cycle and its impact on climate feedbacks. The oceans and the terrestrial biosphere are the two primary reservoirs of carbon in the climate system. Throughout history, these reservoirs have been responsible for absorbing significant quantities of CO_2 emissions caused by human activity. A significant portion, approximately 50%–70%, is eliminated by the oceans, while the remaining amount is absorbed by the terrestrial biosphere. It is important to consider that global warming may have an impact on the ability of these reservoirs to effectively store atmospheric CO_2. The rate of carbon uptake by these reservoirs plays a crucial role in regulating the amount of CO_2 in the atmosphere. If this rate decreases, it could contribute to the accumulation of CO_2 and further exacerbate the problem of greenhouse gas concentrations.

Various paths can be taken by the feedback effect in the world's oceans. Initially, it is important to note that as surface waters increase in temperature, there is a decrease in the amount of dissolved CO_2 they can hold. In addition, an increase in atmospheric CO_2 levels would result in a multiplication of bicarbonate ions (HCO3–) in the oceans, leading to a subsequent rise in ocean acidity. Due to the breakdown of calcium carbonate (CaCO3) by acidic solutions, the increasing acidity poses a threat to marine organisms that rely on CaCO3 for their skeletons or shells. As these organisms struggle to absorb carbon from the ocean, the efficiency of the biological pump that keeps the oceans as a carbon sink would decline accordingly. Additionally, it is important to consider the potential impact of increasing surface temperatures on the thermohaline circulation. This intricate global pattern of oceanic flow plays a significant role in the sinking of surface waters near the poles and the sequestration of carbon in the deep ocean. When there is an increase in the amount of fresh water mixing with saltwater, it can slow down the flow and impact the efficiency of the solubility pump. This pump is responsible for transferring CO_2 from shallow to deeper waters. It is worth noting that if global warming were to persist, there is a possibility that the oceans would transition from being a net sink of CO_2 to becoming a net source (Basics of Climate Change, US EPA, 2024).

Due to the warming and drying of regions like Amazonia, the loss of large sections of tropical forest would result in a decrease in the overall capacity of plants to sequester atmospheric CO_2. Consequently, the terrestrial biosphere, despite its current role as a carbon sink, would transition into a carbon source. One important factor that greatly influences the rate of photosynthesis in plants is the ambient temperature. It is fascinating to observe how various plant species have evolved to thrive in their specific climatic conditions, optimizing their photosynthetic rates. When temperatures rise and conditions go beyond the ideal range for photosynthesis and soil respiration, the rate of photosynthesis will decrease. When plants die and start to decompose, the metabolic activity of microorganisms, which releases CO_2, would gradually surpass the rate of photosynthesis.

In the event of significant global warming, it is worth noting that methane sinks in the oceans and terrestrial biosphere could potentially transition into methane sources. The annual emissions of methane by wetlands can fluctuate, influenced by factors such as temperatures and nutrient input. In certain cases, wetlands may transition from being a source of methane to becoming a sink.

Another important factor to consider is the potential for heightened methane release due to the warming of Arctic permafrost on land and the subsequent release of methane at the continental margins of the oceans, which occurs a few 100 meters below sea level.

At present, the average atmospheric methane concentration stands at 1,750 ppb, which is equivalent to a staggering 3.5 gigatons (3.5 billion tons) of carbon. In the Arctic permafrost, we have a substantial amount of carbon equivalent, estimated to be at least 400 gigatons. Additionally, there is an even more significant amount of carbon equivalent, up to 10,000 gigatons, trapped in a hydrated crystalline form called clathrate on the continental margins of the oceans. There is a belief that a certain portion of the methane that is trapped could potentially become unstable if there is further warming. However, it is important to note that the exact amount and rate of potential emission are still highly uncertain (Climate Change Science, US EPA, 2024).

2.4.3. Water Vapor Feedback

The atmospheric concentration of water vapor is not subject to free variation, unlike that of other greenhouse gases. Rather, it is determined by a physical relationship called the Clausius-Clapeyron equation, which was named after the German physicist Rudolf Clausius and the French engineer Émile Clapeyron, who both lived in the 19th century. This equation takes into account the temperatures of the lower atmosphere and the surface. This relationship suggests that the ability of air to contain more water vapor increases as the temperature of that volume of air rises, presuming that there is a liquid water surface in equilibrium with the atmosphere. This is a reasonable assumption on water-rich oceans but completely incorrect on land. Because of this, the relative humidity (the amount of water vapor in the air as a percentage of its capacity) is close to 100% over oceanic regions and significantly lower (nearly 0% in dry regions) over continental ones. As one might expect, the average relative humidity of the lower atmosphere is about 70%, which is comparable to the percentage of Earth's surface that is covered by the oceans. No matter how hot or cold the Earth gets, this amount should stay pretty much the same. The relative humidity over land areas up to regional scales can be impacted by human land-use modification, such as tropical deforestation and irrigation, which may cause slight changes to the global relative humidity.

With an increase in atmospheric temperature, more water vapor will be present. The net greenhouse effect gets stronger as the surface warms, leading to even greater warming. This is because water vapor is an extremely powerful greenhouse gas, even more so than CO_2. The term for this kind of positive feedback is the "water vapor feedback." It is the main cause of climate sensitivity being much higher than the previously mentioned theoretical value of 0.25°C (0.45°F) for every 1 watt per square meter increase in radiative forcing (Knight, 2023).

2.4.4. Ice Albedo Feedback

The so-called ice albedo feedback is another significant positive feedback on climate. The very nature of ice as a surface—one with a higher albedo—than that of land or water causes this feedback to occur. So, less ice on Earth means less reflectivity, which means more solar radiation gets absorbed, leading to warmer surfaces. At times of relatively extensive global ice cover, like the maximum point of the last ice age about 25,000 years ago, this feedback becomes much more significant. Since there is less ice to melt as the Earth's surface warms, the ice albedo feedback is becoming less important on a global scale (Climate Change Science, US EPA, 2024; Figure 2.4).

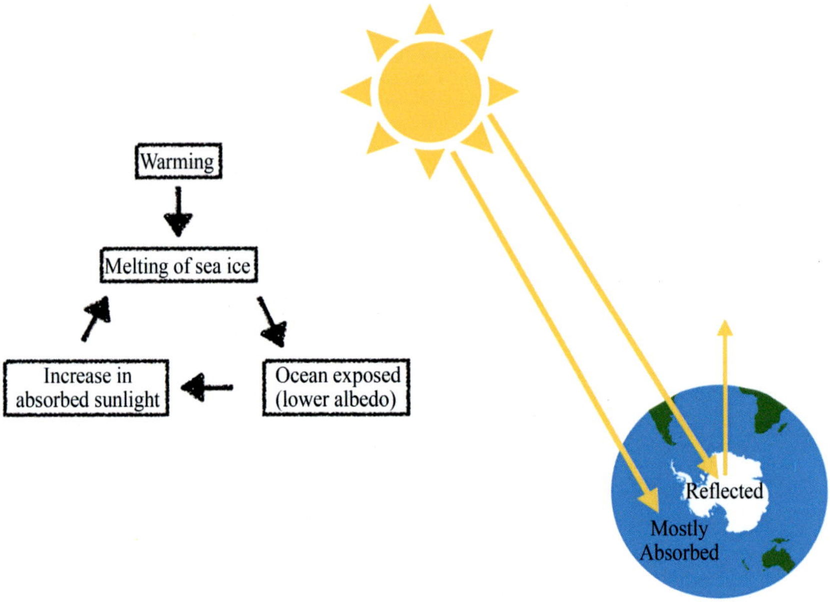

Figure 2.4. *Arctic ocean albedo.*

Source: By Tovarg – Own work, CC BY-SA 4.0, https://commons.wikimedia.org/w/index.php?curid=79657236.

2.5. CLIMATE CHANGE IMPACTS ON ECOSYSTEMS

All the various forms of life that inhabit our planet and the ways in which they interact with the physical world make up ecosystems. Everyday necessities like food, water, air, and building materials like wood and fiber are all provided by ecosystems. In addition to providing a home for a wide variety of creatures, ecosystems serve as venues for human enjoyment, inspiration, and psychological health, as well as buffers against the destructive power of the elements.

Ecosystems are essential to human and environmental health due to their many beneficial characteristics. Nevertheless, ecosystems are facing and adapting to a multitude of stressors caused by humans, with climate change being one of the most pressing issues. Among the many effects of climate change are:

- Higher average temperatures in the atmosphere and water;
- Different patterns of precipitation;
- More intense natural disturbances;

- Changes in the distribution and survival of species; and
- Shifts in the chemical composition of the ocean.

Pollution, alteration of natural ecosystems for human use, movement, and introduction of non-native species, and depletion of natural resources are some of the other stresses that are causing these changes in the environment. All of these things are putting strain on ecosystems, which in turn threatens the survival of species and the ecosystem services that people have grown to depend on.

A lot of people are wondering how vulnerable ecosystems are, how they will change in the future, and whether or not they will be able to continue providing the same services in the future due to the speed of the changes that are happening right now. In light of these worries, scientists are trying to learn more about the impacts of climate change and how ecosystems react to it. They are also trying to find ways to manage ecosystems in a way that makes them more resilient to this threat. Even though climate change is having more and more negative effects, many natural ecosystems have shown remarkable resilience in the face of adversity.

Managing and conserving ecosystems in a way that incorporates both traditional and innovative approaches increases the likelihood of resilience, which occurs in ecosystems that are generally healthy and have maintained a large portion of their original species diversity. When taken as a whole, these methods present a wealth of possibilities for fostering restoration and the ongoing health of ecosystems; however, management decisions involving their implementation may necessitate a thorough evaluation of potential risks, costs, and tradeoffs. In addition to mitigating the effects of climate change, ecosystems can aid in society's adaptation to the changing climate and lessen the severity of disasters caused by it.

By sequestering carbon dioxide, a greenhouse gas that traps heat, in terrestrial biomass and soils or aquatic phytoplankton, algae, and sediments, ecosystems reduce the impact of climate change. The numerous advantages that ecosystems provide also help societies adapt to the effects of climate change. For instance, mangrove swamps and wetlands can shield coastlines from storm surges, erosion, and rising sea levels, while forests can lessen erosion and improve flood control inland. We can learn more about the effects of climate change and ecosystem interactions on society if researchers work to increase our understanding of solutions based on nature, build measurement networks, and bridge the gap between the physical, social, and natural sciences.

Meanwhile, research into ecosystems has progressed to the point where it can guide policy and management choices in light of various prospective futures. To disseminate and expand upon this information, scientists are currently collaborating with various stakeholder groups and decision-makers.

The ecosystems and the people who live in them stand to gain from these endeavors if we can get people to use solutions that are based on nature and follow best practices that improve ecosystem health.

2.5.1. Oceans

There is a continual back-and-forth between the oceans and the air, with each medium exchanging various forms of heat, water, gas, and particle. The ocean takes up some of the heat that is being released into the atmosphere. Both the surface and deep water ocean temperatures are affected by the amount of heat that the ocean stores. From plankton to polar bears, a wide variety of marine life is vulnerable to the effects of ocean warming on their environment and food sources.

Carbon dioxide is absorbed by the oceans as well. When carbon dioxide dissolves in water, it undergoes a chemical reaction that produces carbonic acid. Oceans absorb part of the additional carbon dioxide that humans release into the atmosphere, increasing the amount of carbonic acid in the water. Marine ecosystems, including coral reefs, are vulnerable to the impacts of an acidifying ocean (Knight, 2023).

2.5.2. Forests

While certain forests may experience short-term advantages from a prolonged growing season, it is important to note that climate change is anticipated to contribute to an increase in wildfires due to the extension of the summer fire season. Extended periods of hot weather can place significant strain on trees, rendering them more vulnerable to wildfires, insect infestation, and disease. It is important to note that climate change has had a significant impact on the size and frequency of forest fires, insect outbreaks, and tree deaths, particularly in Alaska and the Western regions. From 1970–1986, the area burned in western U.S. forests was significantly smaller compared to the period from 1987–2003. In fact, during the latter period, the area burned was nearly seven times larger. Over the course of the past three decades, there has been a significant extension of the wildfire season in the Western region, amounting to an additional 78 days (Bernauer, 2013).

2.5.3. Habitats

Plants and animals tend to thrive in habitats that offer them the ideal temperature and precipitation levels. Climate change has the potential to significantly impact and, in certain instances, even devastate various habitats. As an illustration, the melting of sea ice is causing the loss of a crucial habitat for numerous species in the Arctic. Coastal wetlands, such as mangroves, play a vital role in supporting numerous species. However, the alarming rise in sea levels poses a significant threat to their existence. Certain species have the ability to adapt to changing habitats. They can shift their range towards the north or higher altitudes to cope with increasing temperatures. On the other hand, some individuals may struggle to adjust quickly to keep up with the rapid rate of climate change.

2.5.4. Invasive Species

Invasive species are a significant concern in many ecosystems. They can have detrimental effects on native plants and animals, disrupting the delicate balance of the ecosystem. It is crucial to understand the characteristics and behavior of these species to effectively manage and control their spread. As temperature, precipitation, and other conditions fluctuate, the species that are most well-adapted to the new conditions will flourish, often outcompeting others for food and resources. Certain species that flourish could be considered invasive, meaning they are not native to a particular region. As a result, they have the potential to gradually displace or even harm native species.

2.5.5. Migrations and Life Cycle Events

Temperature, humidity, and the quantity of daylight are climatic variables that influence the timing of numerous natural occurrences, including the blooming of flowers and the migration of animals. These regular occurrences could be upended by the unpredictable weather patterns brought on by climate change. Seasonal behavior and interactions among species can be impacted by these disruptions. As an illustration, it is possible that hatchlings won't have enough food if birds migrate and lay eggs at the wrong time. The ability of certain plant and animal species to adapt their life cycles to signals from shifting weather patterns is not universal (Knight, 2023).

2.6. CLIMATE CHANGE IMPACTS ON WEATHER PATTERNS

According to the IPCC, by the end of this century, surface temperatures around the world will have risen by a couple of degrees Celsius. Warming (and changes in other climate variables) will not be uniform across regions and will continue to display inter-annual to decadal variability because of the presence of internal variability and geographic variability in the climate response. It is thus an incredibly challenging task to comprehend climate-driven changes in extreme weather at small spatial scales. More consistent across models and consistent with the first principles of what should happen when more CO□ is released into the atmosphere make some changes, like more days with temperatures of

90°+ F and more days with heavy precipitation, easier to predict. Hurricanes, blizzards, and severe thunderstorms are examples of phenomena whose impacts are notoriously difficult to forecast and which frequently display large variations across model ensembles (Climate Change Science, US EPA, 2024).

The following is a summary of the most recent research on the effects of climate change on the features of several important weather events.

2.6.1. Tropical Cyclones

0The power of tropical cyclones (TCs) comes from the latent heat that is released when water condenses at higher altitudes, which is obtained by the evaporation of water at the ocean surface. The low pressure that results from the Earth's rotation propels cyclonic winds toward the eye, which is located at lower atmospheric levels. Although other factors play a role, the three main requirements for TC formation are as follows:

- Sea surface temperatures (SSTs) that are high enough (>26°C);
- Vertical wind shear, the change in wind velocity with height, must be low enough; and
- The Earth's rotation must be strong enough (>5° N and S) to contribute significantly.

Although TC formation is conceivable year-round, it is most common in the summer when surface ocean waters are warmest and shear is at its weakest. In terms of total TCs and their intensity, the Western North Pacific stands out as the most active ocean basin (Basics of Climate Change, US EPA, 2024).

While short-term weather patterns can have a significant impact on individual TCs, climate-driven changes (such as increasing sea surface temperatures or shifting jet stream positions) impact the statistics of wind and flood hazards related to TCs. These changes would be caused, in part, by increased warming at the poles and the melting of polar ice. These changes have the potential to impact:

- The intensity;
- The size;
- The frequency;
- The seasonality;
- The distribution; and
- The trajectory of TCs.

TCs in different basins may be impacted differently by climate-driven changes, and these changes may vary across basins or smaller scales. The small size, infrequent return period, and high natural variability between years and even decades make it difficult to determine the climate drivers of trends in tropical storms that make landfall (Yusoff and Gabrys, 2011).

At larger scales, trends can be more easily evaluated by utilizing basinwide measures of TC intensity, frequency, and duration, like the Accumulated Cyclone Energy (ACE) or the Power Dissipation Index (PDI). These are defined as the sum of the maximum one-minute sustained wind speeds cubed or squared, at six-hourly intervals, for all periods when the cyclone is at least tropical storm strength.

According to these metrics, TC activity in the North Atlantic has been steadily rising since the 1970s. Over the past few decades, there has been an approximately 25%–30% increase in the proportion of Category 4 and 5 storms for every degree of warming on a global scale. It is still debatable, though, whether or not this trend is due to human interference (Noguer, et al., 2001).

2.6.2. Extratropical Cyclones

The energy for extratropical cyclones (ETCs) comes primarily from the atmospheric gradient in horizontal temperature, as opposed to tropical cyclones, which get most of their energy from the sun. The energy that a developing storm can transform into rotational wind, also known as kinetic energy, is represented by this gradient.

Storms gain kinetic energy when cooler, denser air compresses beneath warmer air, which lowers the center of gravity and causes a decrease in potential energy. Vertical wind shear, which is the presence of the jet stream at higher altitudes and the reason for the density difference across the temperature front, is responsible for the increasing westerly wind speed with height in the mid-latitudes. The fundamental idea behind the 1920s-introduced Norwegian Cyclone Model

is the juxtaposition of air masses of different densities. A process known as baroclinic instability, which was theoretically shown by Charney (1947) and Eady (1947) to occur in the 1940s, allowed extratropical cyclones to form, but only up to a certain threshold of horizontal temperature gradient, or baroclinicity. Additional features and aspects of extratropical cyclone development have been explained by theories since then. It is important to highlight a more recent theory that incorporates an additional relevant process that holds great significance when considering climate change.

According to Shapiro and Keyser (1990), a field study titled ERICA Explosive and Rapid Intensification of Cyclones over the Atlantic, which was carried out in the mid-1980s found that extratropical cyclones can start to exhibit traits seen in tropical cyclones, like a warm core. Like tropical cyclones, they also pointed out that the release of latent heat can have a major impact on development. The surface latent heat flux is determined by the wind speed and the relative humidity of the air 10 meters above and below the Earth's surface (land or water).

When a cold, dry wind blows across a warm, damp surface, the latent heat energy in the surface can be transferred upwards into the atmosphere through a process called flux. From a climate change standpoint, the low to mid-level horizontal temperature gradient and latent heating are important features to consider because their current and future changes will contradict each other, making it more difficult to know how climate change will affect extratropical cyclone activity. In particular, at low to medium altitudes, the north and south poles are expected to warm more than the equatorial regions, according to predictions of future temperature changes. As a result, the ambient temperature gradient from the poles to the equator will be diminished due to this differential warming (Climate Change Science, US EPA, 2024).

2.6.3. Severe Thunderstorms

Mesoscale convective storms that cause damaging hail, wind, or tornadoes are called severe thunderstorms (STs). In the US, a severe thunderstorm is defined as one that produces hail with a diameter of at least 1 inch, gust winds of 50 knots or more within three seconds, or a tornado with an intensity of at least EF-1. To flourish, STs need a particular kind of habitat. To be more specific, convective instability is necessary for two things:

- First, to produce updrafts powerful enough to sustain hailstones in the air until they reach threshold size or larger; and
- Second, to lift moist air parcels from the surface, which cause precipitation to fall as rain or hail.

Intense falling precipitation is needed to generate damaging winds because it must drag air downwards with it with enough force to cause the air to spread out horizontally with damaging velocity when it hits the ground. The destructive wind potential of a thunderstorm can be amplified by these downbursts, which are also known as straight-line winds or derechos (Friman and Strandberg, 2014).

Important for ST formation is the convective available potential energy (CAPE), a vertically integrated measure of atmospheric convective instability. Vertical wind shear is an essential factor in the growth of ST. It refers to the vertical shear of the horizontal wind. Vertical wind shear refers to the variation in wind speed and direction as we move up in the atmosphere. Thunderstorms are able to reach their maximum intensity and sustain it for longer periods because this shear helps to separate the updrafts from the downdrafts (Bernauer, 2013).

Storms can only grow as small as the amount of precipitation that falls through the core updraft region when there is no vertical wind shear. An additional critical component that affects the development of tornadoes is how the shear changes direction with height.

Since convective instability is necessary for ST formation, they usually develop in the summer in both the Northern and Southern Hemispheres. However, the necessary instability can exist during cold seasons in certain places, but at shallower depths. On the other hand, not every area with warm surface air experiences severe storms due to the vertical shear requirement. The weak vertical shear in the tropics, which is a function of the environmental horizontal temperature gradient (baroclinicity), is one reason why severe storms seldom form there.

High CAPE and high shear are the two primary components shared by hail, damaging wind, and tornadoes, though the precise circumstances necessary for each are slightly different. Preferred environments are influenced by geography as well.

Because of its location, the United States is one of the countries most likely to experience extreme weather. Warm, humid, and potentially unstable air is mainly brought to the Great Plains from the Gulf of Mexico. The warm, humid, and unstable air is drawn northwestward by the southeasterly winds that precede low pressure systems that form on the lee side of the Rocky Mountains. Strong south-westerly winds transport considerably drier and cooler air from the Mexican Plateau to higher altitudes. A thermodynamically unstable environment with wind speed and directional shear is produced by this configuration (Friman and Strandberg, 2014).

Also crucial is the manner in which the shear modulates with altitude. Tornadoes can't form, for instance, unless the wind shear profile is veering, or turning clockwise with height in the Northern Hemisphere. While vertical wind shear is presently most noticeable in the middle latitudes and the polar regions, the Earth's surface will warm overall in the 21st century. Surface warming does not necessarily lead to increased thermodynamic instability; however, upper levels, particularly in the middle latitudes, are anticipated to experience cooling or at least less warming as a consequence of climate change. The atmosphere's inherent convective instability will be exacerbated as a result of this scenario (Climate Change Science, US EPA, 2024).

From a global average perspective, vertical wind shear will be weaker because the polar regions are warming more than the equatorial regions. This is similar to what happened with ETCs, where a weaker horizontal temperature gradient was observed. Enhanced instability would encourage more frequent and intense ST activity, whereas reduced shear would have the reverse effect. Again,

GCMs are essential for calculating the overall outcome of the environmental conditions that promote the development of severe storms on a large scale.

2.6.4. Wildfire

The phenomenon known as wildfire is a destructive force of nature that can wreak havoc on the environment and pose significant risks to human life and property. It is a natural disaster that occurs when a fire spreads rapidly across vegetation, fueled by dry conditions and strong winds.

Understanding wildfires involves considering the dynamic interplay between meteorological and hydrological factors, as well as the evolving nature of biomass characteristics.

The meteorological conditions are quite intricate due to the presence of two distinct sets that exhibit some variations in their hydrological characteristics:

1. **Convective Storms and Drought:** Convective storms produce significant amounts of rainfall, which can suppress fires and, of greater significance, they generate lightning, a potential source of ignition for fires. It is not necessary for these storms to be severe, meaning they do not have to generate hail, damaging winds, or tornadoes. Their primary requirement is to produce lightning. Lightning strikes have the ability to start fires, even in areas where the vegetation is not experiencing drought conditions. However, when the biomass is drier, the chances of ignition and fire spread are higher.
2. **Human Activities:** Furthermore, fires can be initiated by various sources, including cigarettes, campfires, arson, and even downed power lines, even when lightning is not present. It is worth noting that a significant majority of wildfires in the United States originate in this manner.

Considering the significant correlation between fire and climate, it is undeniable that climate-driven alterations in fire activity will manifest in numerous regions. However, it is important to note that the relationship between climate change and fire is intricate. It is interesting to note that even a slight temperature change can yield contrasting outcomes (Yusoff, and Gabrys, 2011).

As precipitation levels rise in certain warm grasslands and shrublands, the result is an increase in productivity, more fuel, and subsequently, heightened fire activity during the dry season. In climates with a well-balanced supply of moisture, an increase in precipitation could potentially reduce fire activity (Friman and Strandberg, 2014). Warmer and drier weather can lead to an increase in fire activity in areas with abundant biomass. However, in moisture-stressed environments, this type of weather can have the opposite effect.

Increased evaporation reduces the growth of biomass, which is essential for fueling fires. Comprehending the potential impact of climate change on regional climate variables such as temperature, precipitation, lightning, drought, surface wind speed and direction, and biomass growth is a complex task that goes beyond relying solely on the results of a GCM. One common method involves utilizing data generated by GCMs and feeding it into DGVMs. The DGVMs simulate the climate-based processes that govern plant growth and mortality in various vegetation types. Many of these models have integrated a fire module, as demonstrated by previous studies (Noguer et al., 2001; Figure 2.5).

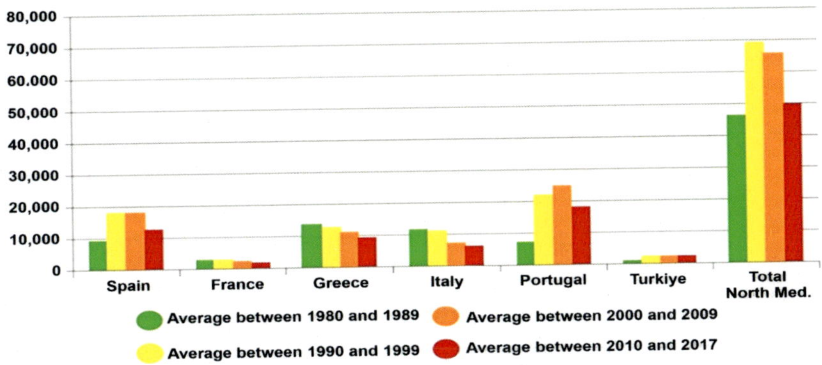

Figure 2.5. *Climate change is raising global temperatures and drier conditions, increasing wildfire risk.*

Source: Farid A, Alam MK, Goli VSNS, Akin ID, Akinleye T, Chen X, Cheng Q, Cleall P, Cuomo S, Foresta V, et al., A Review of the Occurrence and Causes for Wildfires and Their Impacts on the Geoenvironment. Fire. *2024; 7(8):295. https://doi.org/10.3390/fire7080295.*

2.6.5. Heavy Precipitation and Inland Flooding

The predicted effects of climate change on flooding can be better understood by looking at the climatological factors that influence the intensity of various floods. The most challenging type of flooding is likely to be fluvial or riverine flooding, which necessitates knowledge of flood management techniques and the potential behavior of weather patterns over long time periods (e.g., weeks, months, or years), like the ones that occurred in the Midwest in 1993. Perhaps more easily understood is pluvial, or rain-related, flooding, which mainly requires understanding the potential future changes in heavy precipitation events. Storm surges caused by winds from both tropical and extratropical systems are more closely linked to coastal flooding than the characteristics of precipitation.

Thermodynamically, the relationship between a warming atmosphere and changes in precipitation is relatively simple, in contrast to the analyses of the effects of climate change on particular weather systems like tropical cyclones, extratropical storms, and severe thunderstorms. The saturation vapor pressure of water grows exponentially with increasing temperature. A more basic explanation is that air can hold more moisture when it is warmer. However, this is not totally true since water vapor can exist even when there is no air present. The expected outcome of climate change is that precipitation events will become more intense, if not more frequent because more of the moister (absolute) air condenses as it rises (Basics of Climate Change, US EPA, 2024).

By utilizing ocean salinity as an additional integrated metric for long-term alterations in the water cycle, Durack et al. (2012) demonstrated, in a novel manner, the fundamental impact of temperature on precipitation. An intensification of the global water cycle of 8+/-5% per degree of surface warming was revealed by their findings. The substantial spatial heterogeneity of precipitation can make it difficult to find direct evidence of changes.

The historical record provides relatively strong evidence that quantities related to precipitation, including heavy precipitation events, have been on the rise, despite that complexity. Yusoff and Gabrys, (2011) discovered that while precipitation was trending upwards globally and in sub-polar and equatorial regions, it was trending downwards in the subtropics. Precipitation over mid-latitude land areas in the Northern Hemisphere has also increased, according to the 2013 IPCC report. The average yearly precipitation in many parts of the world, including the US, ESA, Europe, India, NW Asia, and South-eastern China, has been on the rise recently (Climate Change Science, US EPA, 2024).

2.6.6. Sea Level Rise and Coastal Flooding

While the situation with freshwater floods is more murky, coastal floods are becoming more frequent and more severe (Noguer, et al., 2001). For instance, the average number of days that New York City experiences flooding has increased from 19 in 1920 to 70–99 now. The available evidence indicates that, in most cases, the increased frequency of extreme coastal flooding events is primarily caused by the rise in mean sea level (Church et al., 2013). According to Noguer et al. (2001), the majority of the 1.5 mm/yr rise in sea level since 1900 can be attributed to global warming. This rise is roughly split 50/50 between the expansion of ocean water due to heat and the melting of continental ice sheets and glaciers.

Every method of reconstruction indicates that the rate of sea level rise has recently accelerated to over 3 mm/year, which is within the last 20 years. Supporting the common belief that melting land ice is a major contributor to this acceleration, satellite-derived observations of glaciers and ice sheets showing thinning, acceleration, and mass change corroborate this (Noguer et al., 2001).

Variations in weather patterns, ocean circulation, and processes unrelated to climate change, like land subsidence, cause sea levels to vary significantly across regions. Subsidence, also known as vertical land motion (VLM), is caused by tectonic activity, soft-sediment compaction from overburden or the withdrawal of groundwater or hydrocarbons, and deformation from the ice-ocean mass transfer (Yusoff and Gabrys, 2011).

Land subsidence is the main cause of relative sea level rise along the U.S. Gulf Coast and in certain Asian megacities.

The coast of the Northeast and Mid-Atlantic states in the United States is another region where sea level has risen at a rate two to four times faster than the world average of 2.3 mm/year over the past 30 years. According to Noguer et al. (2001), the New York City region saw a 128 mm increase in 2009 and 2010. This increase was partially caused by a negative North Atlantic Oscillation and a slowing of the Gulf Stream.

Sea levels in the western tropical Pacific have fluctuated at a rate of about 10 mm/year due to the action of winds and currents. Some trends were seen to have moderated or reversed during the satellite period, and reanalysis products provide strong proof of long-term climate oscillations. Nonetheless, secular, human-caused climate change may be responsible for part of the increase in sea levels in these areas.

Coastal floods are predicted to become much more common as a result of ongoing climate change. Reasons for the strong projections of an increase include the fact that, in the long term, regional sea level will rise at most global locations due to the thermal expansion of ocean water, ice melting, and changes in terrestrial water storage, while there may be periods when it falls. Based on the predictions made by GCMs, the warming and expansion of the oceans will likely persist and expand to deeper depths at a rate related to atmospheric feedbacks and the ocean's heat uptake (diffusivity) (Kuhlbrodt and Gregory, 2012). At least some of the land ice, or glaciers, will likely melt away in the coming century, with the melting of the Arctic and Alaska glaciers, as well as those on the periphery of Greenland and Antarctica, expected to have the greatest impact on global sea levels (Bernauer, 2013).

Lastly, various other factors, such as land-based storage (such as reservoirs and groundwater withdrawal), could potentially impact sea level; however, it is generally believed that these factors are less substantial and carry less uncertainty. Forecasts need to take into consideration VLM changes that might be caused by human actions (like fluid withdrawals) in a lot of places.

2.7. CLIMATE CHANGE MITIGATION AND ADAPTATION STRATEGIES

Climate change mitigation and adaptation activities are crucial in addressing the challenges posed by climate change. These activities play a significant role in combating the impacts of climate change and ensuring a sustainable future. By mitigating climate change, we can reduce greenhouse gas emissions and limit the extent of global warming. Additionally, adaptation activities help communities and ecosystems adapt to the changing climate, minimizing the negative consequences. Together, these efforts contribute to a comprehensive approach to tackling climate change.

A worldwide decrease of 7.6% in GHG emissions per year from 2020–2030 is necessary to meet the goal of the Paris Agreement, which is to keep global warming well below 2°C–1.5°C, relative to pre-industrial levels. Raija Laiho, a Research Professor at the Natural Resources Institute Finland LUKE, emphasizes the crucial importance of climate change mitigation. According to Friman and Strandberg (2014), it is essential to limit the ongoing change to a manageable level in order to prevent the risk of a complete and uncontrollable collapse of ecosystems, including agriculture and forestry.

The IPCC emphasizes the importance of active mitigation efforts in six key sectors:

- Energy;
- Industry;
- Agriculture and Food;
- Forests and Land Use;
- Transport; and
- Buildings and Cities.

Regardless of the potential success of mitigation efforts in reducing emissions, it is important to acknowledge that the warming effects of the atmosphere have already had and will continue to have a significant impact on our society's livelihood. Climate change adaptation strategies are crucial in addressing this issue. According to Friman and Strandberg (2014), adaptation refers to the various measures we can implement to effectively deal with the repercussions of continuous change. The amount of adaptation required is contingent upon the extent and pace at which we can mitigate. According to the European Commission, our society will need to make significant adjustments in various sectors such as:

- Infrastructure and Buildings;
- Energy;

- Agriculture and Forestry; and
- Insurance, and Tourism.

Adaptation is a crucial aspect that has garnered recognition at the EU level. This was evident on February 24, 2021, when the European Commission adopted a new strategy for climate change adaptation. The ultimate objective is to achieve climate resilience by 2050. The strategy emphasizes the significance of making data-driven decisions when implementing adaptation measures. It is clear that the agricultural and food industry must address and adjust to the challenges posed by climate change. In addition, it is important to consider the specific context of each sector and location when addressing climate change adaptation and mitigation.

According to Kaido Soosaar, an Associate Professor of Environmental Technology at the University of Tartu, the significance of mitigation or adaptation in relation to climate change varies greatly depending on the scale and sector of activity. Although land use and land use planning at the local scale can effectively contribute to mitigating climate change, it is important to recognize that adaptation is still necessary. It is crucial for land use at a specific site and during a specific period to adapt to changing environmental conditions.

2.7.1. Possible change in Agriculture, Forestry, and Other Land Use (AFOLU)

The AFOLU sector—namely, agriculture, forestry, and other land uses—requires substantial adaptation and mitigation efforts. Between 2007 and 2016, the AFOLU sector was responsible for 23% of all net anthropogenic greenhouse gas emissions, according to the IPCC. The European Commission approved a package of laws in July 2021 that would regulate forestry and land use from 2021–2030. By 2030, the EU aims to have reduced its greenhouse gas emissions by 55% according to the proposals (Basics of Climate Change, US EPA, 2024).

The "no debit" rule is fundamental to the proposals; that is, the sector's CO_2 removal efforts should equalize all land use-related emissions. As an example, planting seven hectares of forest would have to be planted to make up for one hectare of lost peatland. This is just one example of how the land use sector makes emission compensation complicated. There isn't enough room to plant enough trees without causing conflicts with other land use interests, so this kind of emission compensation could be difficult in Europe (Friman and Strandberg, 2014).

The AFOLU sector has the potential to make a significant impact on adaptation and mitigation initiatives on a global and regional scale. The sector's significance has been acknowledged at the European level, and its mitigation efforts are anticipated to revolve around regulations that promote sustainable forestry and agriculture. Droughts are predicted to affect the agricultural sector more often and with more severity as a result of a lack of climate mitigation efforts aimed at reducing potential global warming levels.

The IPCC has identified many drying regions around the world, including parts of Western and Central Europe, where this trend is anticipated. This sets in motion a cascade of events that increases emissions of greenhouse gases, such as the frequency and severity of forest and peatland fires and the melting of permafrost soil. In contrast to droughts, which are becoming more common as a result of climate change, floods, increased soil erosion, nutrient losses, and eutrophication of rural waterbodies are all consequences of warmer air masses bringing intense precipitation events in the summer.

More drought-tolerant crop species, new varieties with a longer growing season, and winter crops that harden and survive in situations with unstable or no snow cover are all necessary adaptations to climate change, even though maintaining or increasing carbon sequestration is the preferred mitigation strategy.

2.7.2. Gradual Actions Made to Achieve Significant Objectives

Taking smaller steps is crucial when it comes to achieving our big aims, such as improving the national inventory of greenhouse gas emissions. Understanding the dual role of the AFOLU sector in the carbon cycle is essential. Soils and forests can both release and absorb carbon, making it vital to promote practices that align with our climate goals. Within this context, Kaido Soosaar highlights a significant concern regarding carbon accounting. He emphasizes that as long as consumption and demand remain unchanged, emissions are unlikely to decrease. Instead, they may be substituted by other sources or activities that are not thoroughly examined in terms of emission counting, primarily due to disregarded external factors. For instance, the substitution of peat products with coconut residues may not be adequately studied in terms of its impact on emissions. It is of utmost importance to implement strategies that effectively reduce greenhouse gas emissions without simply transferring them to another sector or location. To achieve this, it is crucial to have a robust system in place for accurately monitoring and accounting for emissions based on scientific principles.

The calculation of emissions is a highly intricate task to improve national GHG inventories. National GHG inventories play a vital role in evaluating and monitoring progress towards GHG reduction targets based on data. On the other hand, the national inventories rely on estimations for GHG emission factors for each action in the sectors. As more and better data becomes accessible, the accuracy of the factors and estimates will improve. AFOLU is a crucial sector that national inventories are required to report on. The aim should be to address the knowledge gaps surrounding GHG emission assessment and reporting in nutrient-rich drained organic soils. Through the project, national inventories will be enhanced by expanding and verifying GHG emission factors and activity data in cropland, grassland, and forest land (Climate Change Science, US EPA, 2024).

CHAPTER 3

NATIONAL SECURITY IMPLICATIONS OF CLIMATE CHANGE

Contents

INTRODUCTION

The effects of climate change are contributing to an increase in natural disasters, refugee flows, and conflicts over fundamental resources such as food and water. This is a threat to our national security that is both urgent and growing stronger. All the way from the Arctic to the Midwest, people are beginning to experience the effects of climate change in reality. Storm surges and rising sea levels pose a threat to coastal areas, as well as to public infrastructure and private property. As a result, the economy of the entire world is negatively impacted, which also contributes to the rising costs of preparing and restoring infrastructure.

3.1. CLIMATE CHANGE AND CONFLICT

Researchers have been studying the intersection of climate change and conflict since the 1970s. Although climate change alone is not considered a direct cause of violent conflict, it is important to recognize the intricate relationship between the two phenomena, which can have devastating consequences.

According to a recent study conducted by UNOCHA, it has been found that countries experiencing high levels of violence are at greater risk when it comes to the impacts of climate change. According to the Notre Dame Global Adaptation Initiative Index, it is confirmed: It is worth noting that Afghanistan, the Democratic Republic of Congo, Mali, Somalia, and Yemen, all find themselves in a state of violent and protracted conflict, making them particularly vulnerable to the impacts of climate change.

Figure 3.1. *Climate change is seen as a global security and stability threat.*

Source: By Eddiem360 – Own work, CC BY-SA 4.0, https://commons.wikimedia.org/w/index.php?curid=94018959.

Climate change has the potential to significantly exacerbate conflicts, and in turn, conflicts can also worsen the impacts of climate change (American

Security Project-ASP, 2023). There are several ways in which climate change can have a significant impact on conflict. It is important to understand these influences in order to grasp the complex relationship between climate change and conflict (Figure 3.1).

3.1.1. People Lose Income Due to Climate Change

One of the consequences of the climate crisis is the negative impact it has on people's income. Several studies have shown that there is a connection between climate change and conflicts, particularly in communities that are involved in farming or herding.

This disintegration occurs because pastoralists and farmers rely heavily on stable ecosystems to support themselves and their families. Their harvests are reduced, their livestock is killed, and their incomes are cut in half due to drought, floods, land degradation, and resource exploitation. As a result of people's inability to cope, the social contract may collapse if this pattern persists over an extended period of time. Because of this, violent conflict is more likely to occur.

For instance, according to the United Nations, economic difficulties caused by the climate crisis "lowers the threshold" for individuals to participate in violent acts in Somalia.

It is quite clear that the modern-day South Sudan is a prime example of this. A flood in 1991 wiped out livestock and crops in the country's Upper Nile region. Cattle raiding was a typical response among the resource-depleted pastoralist Dinka and Nuer people. Unfortunately, the Dinka-Nuer conflict resulted in even more devastation, including the deaths of more people and the destruction of even more communities, despite their initial optimism about the opportunity to recover from the flood (American Security Project-ASP, 2023).

3.1.2. Climate Change and Increased Migration

It becomes evident that when your livelihood depends on the land and it fails to support you, the only viable option is to search for alternative land. However, one major concern that arises is the issue of scale. Climate extremes have led to the unfortunate situation where countless individuals across the globe have been let down by their land, compelling them to make the challenging choice of leaving their homes in pursuit of better opportunities (Figure 3.2). Finding resources has become increasingly challenging. The situation becomes increasingly tense as migrants and locals find themselves in competition for land and water resources. When race or ethnicity are involved, these tensions can be heightened. In the 1980s and 1990s, there was a significant migration in Sudan as people sought out regions with more abundant land resources. An escalation in violence ensued.

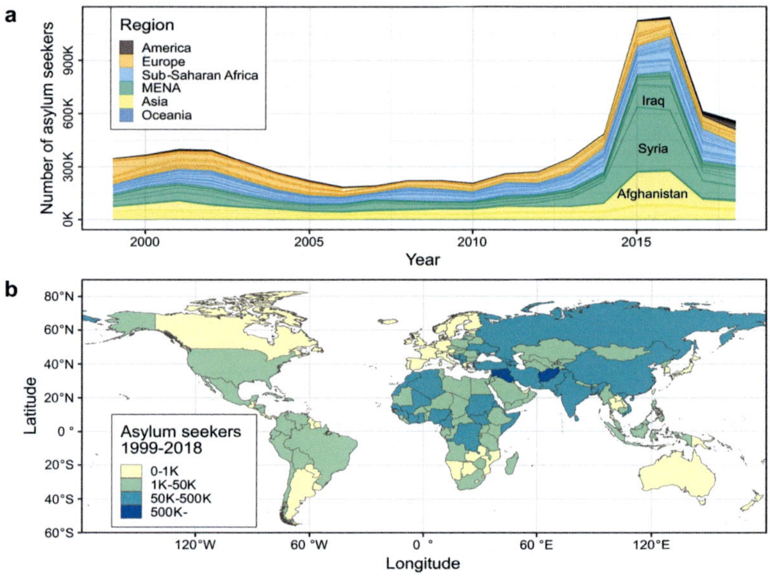

Figure 3.2. *Climate change plays a significant role in the rise of migration patterns.*

Source: Schutte, S., Vestby, J., Carling, J. et al., Climatic conditions are weak predictors of asylum migration. Nat Commun 12, 2067 (2021). https://doi.org/10.1038/s41467-021-22255-4.

3.1.3. Powerful People Can Exploit Natural Resources and Local Conflicts

The conflicts surrounding natural resources can have severe and fatal consequences. However, it's worth noting that these organizations tend to have a more localized focus, as they typically operate within a specific region of a country. Nevertheless, these conflicts can serve as effective tools for local and national elites to divert attention from their own failures and garner support by promoting an "us vs. them" mindset. The escalation of a localized conflict over resources can quickly transform into a nationwide concern, accompanied by a surge in violence.

3.1.4. Climate Change Can Increase Political Instability

One of the many underlying factors contributing to the Syrian conflict is the weather. The nation was hit by a multi-year drought that started in the middle of the 2000s, brought on by unpredictable weather patterns. Along with this, there were severe water shortages caused by antiquated irrigation systems and inefficient watershed management. More than one million Syrians lost their jobs, became food insecure, and had to flee their homes because water became a luxury. In the long run, this exacerbated the already existing political unrest.

3.1.5. Climate Change Hinders Peacekeeping

One important factor to consider is the impact of extreme weather events and climate disasters on peace and conflict. These events have the potential to hinder peace efforts and prolong existing conflicts. Climate-related environmental change has been identified by the United Nations Assistance Mission in Somalia as a significant obstacle to its peacekeeping efforts in the country. This exacerbates three of the challenges responsible for the crisis in Somalia:

- Farmer-herder conflicts;
- Insurgent group recruitment; and
- Competition over land.

Although it may appear to be a chicken-egg scenario, the ultimate outcome remains unchanged: an extended period of violence that impacts a greater number of individuals as time passes (UNEP, 2024).

3.2. CLIMATE CHANGE, DISPLACEMENT, AND MIGRATION

As of 2015, the United Nations (UN) reported that around 244 million people were residing outside their country of birth, while an additional 740 million had either been internally displaced or relocated within their own country (United Nations, 2007). Determining the exact extent to which this movement is attributable to climate-related hazards, and the degree to which these hazards were impacted by climate change, poses a significant challenge. Throughout history, storms, floods, and droughts have been a part of our natural environment. While it is clear that climate change is impacting the characteristics and patterns of these climate-related hazards, it is important to note that not every hazard can be solely attributed to or influenced by climate change (Wilson Center, 2014). It is worth noting that not all countries have a consistent practice of gathering data on immigration or keeping track of the numbers of refugees and internally displaced people. Furthermore, even among countries that do collect such data, there are significant variations in the methods used for collection and reporting (World Economic Forum, 2023).

There are certain challenges that make it difficult to accurately track the movement of cross-border migrants and refugees. These challenges include:

- Double-counting;
- Slow asylum registration processes; and
- Cumulative counting of international migrants in some datasets.

These factors can complicate efforts to understand migration patterns over time.

Various organizations such as the IDMC, the Red Cross/Red Crescent, and the UN Refugee Agency (UNHCR) gather data on an annual basis. However, it is important to note that their methodologies and datasets may not always be directly comparable, both among themselves and with the statistics maintained by individual countries.

Tracking internal movement poses challenges that are closely tied to factors such as the timing of displacement, whether individuals have been displaced before, or if they are involved in cyclical migration. In addition, it is important to note that not all climate-related hazard events result in disasters. Therefore, datasets that specifically track disaster displacement may not capture all instances of climate-related hazard displacement. This highlights the challenge in categorizing and monitoring human mobility when it comes to hazards versus disasters (American Security Project-ASP, 2023).

Currently, there are no datasets available that specifically document the displacement of individuals caused by droughts or other gradual hazards, even when these events escalate into full-blown disasters. Datasets do not fully capture the intricate relationship between environmental degradation, human mobility, and the impact of climate change on environmental degradation. When examining the statistics, it becomes clear that tracking international movement does not offer much insight into the impact of hazards that occur gradually or unexpectedly.

On a global scale, the movement of people across borders has shown a consistent pattern from 1990–2010. According to UN data, which considers international migrants as individuals living outside their country of birth, including refugees, long-term displaced individuals, and voluntary migrants, the numbers have remained relatively stable. The figures have ranged from 41.4 million (1990–95) to 41.5 million (2015–20).

When we examine cross-border migration and refugee flows from 1965–2015, we find that the total number of movements has seen a modest increase in proportion to the overall growth of the population. Specifically, the percentage has risen from 2.5% in 1965–3.3% in 2015. The number of refugees has remained relatively stable, with 20.6 million in 1992 and 21.3 million in 2015 (American Security Project-ASP, 2023). It is worth noting that, apart from Latin America, most individuals who migrate tend to stay within their own region of origin.

Unfortunately, without sufficient data, it is not feasible to establish connections between cross-border movement and climate hazards that occur suddenly or gradually. It is worth noting that only internal displacement databases keep records of those who have been displaced by sudden-onset hazards. One of the challenges that arises is the need for evidence-based policy and decision-making to provide adequate support for the most vulnerable individuals.

When considering human mobility within countries, it is worth noting that the Global Knowledge Partnership on Migration and Development (KNOMAD) project has made estimations regarding internal migration (people moving within their own country) and suggests that the number likely exceeds 700 million individuals annually. However, it is important to acknowledge that the exact scale of internal migration remains uncertain.

According to the IDMC (2017) report, the total number of internally displaced individuals, taking into account both conflict and sudden-onset hazards, reached 40.3 million in 2016. According to the source, the Democratic Republic of Congo, Syria, Iraq, Afghanistan, and Nigeria were the top five countries with the highest numbers of internally displaced persons in 2016.

Internal displacement linked to sudden-onset climate hazards is more prevalent in areas with high vulnerability and limited ability to cope on-site. In certain countries like China and the United States, climate-related hazards have the potential to cause significant displacements. This is partly due to the presence of effective early warning systems and evacuation protocols.

Following environmental stressors, such as climate-related hazards, individuals often choose to remain in their current location to preserve their social networks and livelihoods, as relocation can be a burdensome endeavor. When impacts become too much for someone to handle, they may find themselves needing to move, possibly even through evacuation.

In cases where there is enough capacity, individuals may choose to migrate temporarily or cyclically as a means of mitigating or adapting to certain climate-related hazards. When hazards take a longer time to develop, individuals have the opportunity to adjust their strategies and resources. This can include adapting to their current location or building stronger social and financial support systems. This flexibility provides more options when it comes to deciding whether and where to migrate (State of the Planet, 2023).

Research conducted in Vietnam indicates that when faced with sudden-onset events such as floods or typhoons, individuals are more likely to migrate internally, either temporarily or permanently. On the other hand, people tend to choose to adapt to slow-onset events in their current location (Wilson Center, 2014). In Indonesia and Bangladesh, individuals are choosing to temporarily relocate in response to extreme weather events such as flooding. However, they are opting for longer-term migration when faced with persistent environmental stress and repeated slow-onset events like prolonged heat and drought during crucial agricultural seasons (United Nations, 2007).

In situations of conflict and unstable governance, the occurrence of a climate-related hazard or multiple climate hazards can worsen tensions and lead to prolonged displacement or internal migration to urban areas. There are cases where individuals are forced to relocate for extended periods of time after a disaster due to their vulnerability and the insufficient responses they receive (UNEP , 2024). Instances of prolonged displacement and inadequate government relief systems can be observed in various flood events. For instance, the 2016 floods in Louisiana resulted in the displacement of numerous households for several months, as their homes and assets were severely damaged. Similarly, the 2008 Kosi floods in Nepal and Bihar, India, left many displaced individuals living on embankments for up to a year due to the slow response and recovery efforts. Climate has the potential to significantly impact patterns of migration and displacement due to the heightened frequency and severity of extreme weather events and alterations to seasonal patterns. The impact of climate change on human mobility extends beyond the immediate hazard. It is influenced by factors such as vulnerability and capacity, which shape the choices individuals have to either relocate or remain in place when faced with climate-related hazards. Furthermore, while these hazards may sometimes require people to move, they can also limit their ability to do so. The relationship between hazard and mobility can often be complex and difficult to determine, even with access to data. In some cases, this connection may remain unclear, despite the availability of information. When considering the impact of human mobility in the face of a changing climate, it is important to acknowledge the intricate web of factors at play. As a result, it is currently impossible to provide definitive projections of the positive or negative consequences for countries. Furthermore, it is uncertain whether such projections will ever be possible. The role of underlying vulnerability and capacity drivers, along with the policies and practices governing formal and informal support, highlight that the impact of climate change on human mobility is a matter of sustainable development. It goes beyond just disaster risk management or humanitarian assistance and requires attention accordingly (American Security Project-ASP, 2023).

3.3. CLIMATE CHANGE AND FOOD SECURITY

Food security is based on three pillars:

- Food availability;
- Food accessibility; and
- Food utilization.

Food access is defined as the capacity to obtain adequate quality and quantity of food to fulfill the nutritional needs of all members of a household, as opposed to food availability, which is defined as the presence of consumable food stocks. How food is utilized, the amount of nutrients it contains, and the ability to maintain a balanced diet are all factors that affect food utilization. For food security to be achieved, these three components of the food system must all be satisfied.

According to World Economic Forum (2023), producing enough food to provide a basic but sufficient diet to every person will require a substantial increase in global food production during the 21st century. If the present patterns in eating habits and food system management persist, the demand for food will only increase. Affordability of nutrition, promotion of healthy eating habits, efficient use of resources, development of low-waste supply chains, and resilience to climate change should all be priorities at the same time. All of these things put together will make a food system that can last.

Avoiding further acceleration of climate change and threats to the long-term viability of global agriculture requires a concerted effort to reduce greenhouse gas emissions from agriculture in tandem with intensifying food production. While it is technically possible to make these changes, immediate, global, and significantly increased action is required (Berhane, 2018).

Adaptation by rural households to climate variability and droughts has been going on for decades places around the world. In extremely dry regions, households have even gone "beyond climate" dependence (State of the Planet, 2023). Farmers transition from cropping to livestock holding during dry spells. For instance, studies have shown that farmers in Mali are more likely to switch to a shorter-cycle sorghum variety from the traditional one when the rainy season is shorter. On the other hand, it has been discovered that rain-fed cereal production is falling as a result of climate change, increased reliance on migration, livestock, small-scale commerce, and gardens, and a general decline in agricultural output (State of the Planet, 2023).

3.3.1. How Changing Weather Conditions Affect Food Supply

The living conditions of farmers, fishers, and forest-dependent people who are already vulnerable and food insecure will be further exacerbated by climate change. There is a concerning trend of rising hunger and malnutrition. Rural communities, especially those residing in vulnerable environments, are confronted with an urgent and escalating threat of heightened crop failure, livestock loss, and diminished access to marine, aquaculture, and forest resources. Increased frequency and severity of extreme weather events will negatively affect food availability, accessibility, stability, and utilization, as well as livelihood assets and opportunities in both rural and urban areas. Economically disadvantaged individuals may face the threat of food insecurity as a result of asset depletion and insufficient insurance protection (Figure 3.3).

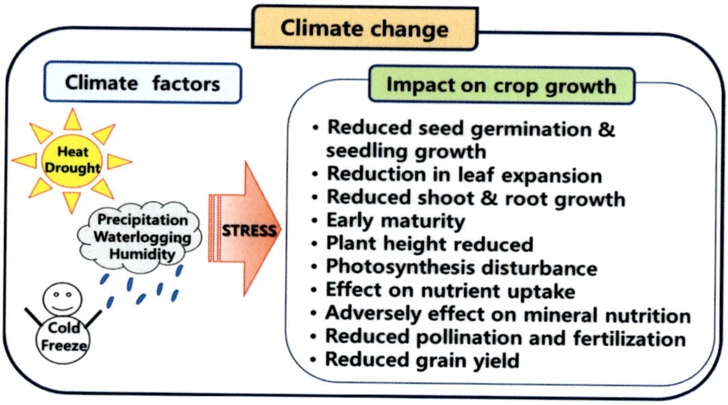

Figure 3.3. *The effects of drought.*

Source: Kim K-H, Lee B-M. Effects of Climate Change and Drought Tolerance on Maize Growth. Plants. *2023; 12(20):3548. https://doi.org/10.3390/plants12203548.*

The ability of rural communities to adapt to the impacts of climate change is influenced by various factors. These include the cultural and policy context, as well as socio-economic aspects such as gender, household composition, age, and the distribution of household assets. In a study conducted by Dahal et al. (2022), the impact of a 4°C temperature rise on major crops in various regions of Nepal was examined. The findings revealed that in mountain regions, the yields of rice, wheat, and maize are projected to increase by 36%, 33%, and 27% respectively. However, in plain regions, there is a slight decrease in the production of these crops. According to Dahal et al. (2022), a general warming would have a positive impact on agriculture in hilly and mountainous regions.

Berhane, (2018) argues that agricultural production is significantly impacted by climate variability. He assert that agriculture, being highly sensitive to climate conditions, is particularly vulnerable to the risks and impacts of global climate change. Except for certain areas in Africa with mid- to high-latitudes, where crop productivity might see a positive impact due to climate change, there is a projected decrease in mean crop productivity across Sub-Saharan Africa. It is widely anticipated that the impact of climate change on food availability in Sub-Saharan Africa will be quite significant. One of the main reasons for this is the vulnerability of subsistence farmers, who are thought to have a limited ability to deal with environmental stressors (American Security Project-ASP, 2023).

The combination of warming temperatures, abnormal rainfall deficits, and more frequent high-intensity rain events can lead to drought, soil degradation, and desertification. Certain crops may exhibit greater resilience in unfavorable conditions, leading to a reduction in crop diversity and consequently a more limited range of food sources. This can have significant implications for human health.

Increased volatility of rainfall can have significant impacts on food crop yields, which makes subsistence agricultural systems particularly vulnerable. Climate change is expected to limit the accessibility of irrigation, as the expenses associated with constructing and upkeeping the technology often outweigh the potential advantages. The relationship between water insecurity and food insecurity is undeniable. As climate change continues to put pressure on water resources, the consequences for crop productivity will be substantial. Preventing disruption or decline in global and local food supplies caused by climate change requires implementing more efficient irrigation and watershed management, enhancing crop varieties, improving land cultivation practices, optimizing farm and livestock management, and developing crop varieties and breeds that can thrive in changing climatic conditions. Understanding climate data and forecasts is crucial for assessing the effects of climate change on agricultural production and the entire food chain.

3.3.2. Climate Variability: The Effect on Food Access

Access to food is determined by the household's capacity to acquire food through various means such as:

- Purchasing;
- Gathering;
- Current production; and
- Relying on existing stocks.

Additionally, food can also be obtained through transfers from relatives, community members, government programs, or donations. The distribution of resources within a household plays a crucial role in ensuring food security for all members of the household. The availability of food in the market, market prices, productive inputs, and credit all play a role in influencing food access (American Security Project-ASP, 2023). Insufficient market infrastructure and an unfavorable policy environment can result in elevated and fluctuating prices for food and inputs, which in turn can have a detrimental impact on agricultural productivity, food supplies, and derived incomes.

Access is contingent upon a variety of factors, including physical, social, and economic considerations. Transport systems in many developing countries can be inefficient and ineffective, which can have a negative impact on the delivery and pricing of food. According to the IPCC (2001), climate change is anticipated to exert pressure on transport systems. The rising food prices and unemployment could potentially worsen people's ability to access them. It is anticipated that the rise in population will also lead to a surge in food prices due to the heightened demand and the escalating costs of imports. The combination of limited financial resources and the anticipation of increased food prices will have a significant impact on accessibility. This, in turn, could potentially contribute to food insecurity, especially in light of the challenges posed by climate change.

3.3.3. Climate Variability and Food Use

Proper utilization of food is achieved through:

- The correct use of food;
- Employing appropriate techniques for food processing and storage;
- Possessing and applying sufficient knowledge of nutrition and child care techniques; and
- Having access to adequate health and sanitation services.

The concept of food utility revolves around the utilization of food. One aspect to consider is the frequency and composition of meals. There are several factors that can limit the utilization of food. These include:

- The loss of nutrients that occur during food processing;
- Insufficient sanitation practices;
- Improper care and storage of food; and
- Cultural habits that can have a negative impact on the consumption of nutritious foods by certain family members.

Food utility can vary with the seasons and availability can fluctuate throughout the year in regions where food is produced and consumed locally (United Nations, 2007).

The hungry season refers to the period prior to the maturation of the planted crops, when they are not yet suitable for consumption. During the harvest season, it is common to witness vibrant festivals and a significant increase in food consumption. In situations where there is a drought and food availability is limited; it is common for the range of food options to decrease. As a result, the frequency of meals may decrease and the balance of nutrients can become inadequate. Malnutrition in children can be a potential consequence of this.

It is worth mentioning that climate can indirectly affect food utility. Households often experience fluctuations in both cash and in-kind income, which can vary within a single year and from year to year. Furthermore, it is important to consider the cultural values, attitudes, and beliefs that influence food choices and eating habits when making changes to one's diet. In cases where the production and consumption of certain foods are intertwined with culture, it is important to consider the potential impact of any changes (World Economic Forum, 2023).

3.3.4. Dealing with Rainfall and Temperature Change

In an effort to lower their emissions of greenhouse gases, the United States and the European Union have imposed strict regulations. The regulation of transport between the agricultural product-producing countries and the purchasing states has been instrumental in this. This is being done to lessen the impact on the environment by cutting down on emissions of greenhouse gases.

Inspiring people to eat seasonal, locally grown food reduces the need to import food, which in turn helps cut down on emissions of greenhouse gases (Wilson Center, 2014). This occurs because energy is wasted when irrigated crops are transported and grown during the off-season.

Investing in systems that collect and efficiently use all precipitation is a certain way to lessen the impact of less rainfall and higher temperatures (World Economic Forum, 2023). This might be a sign of good things to come since it would guarantee that communities will have an ample supply of food grown using the water that has been collected.

After aiming to provide food security to the community, it is equally necessary to use innovations to prevent losses caused by post-harvest handling. As a backup plan for developing economies facing rainfall and temperature variability, the ability to offer insurance programs to poor smallholder farmers and subside in the event of crop failure has also been confirmed (UNEP, 2024).

China has implemented a commendable policy initiative to safeguard its vulnerable population from the impacts of climate change. One of the key measures taken by the country is the significant investment made in irrigation programs (American Security Project-ASP, 2023).

The ability to access irrigation facilities at commercial points has been guaranteed to the communities through government programs. As a result, local communities can rest assured that they will have food on the table. Additionally, communities in Bangladesh and Mexico have also reaped the benefits of government-built water pans and irrigation canals. To ensure the population's food security in a sustainable way, multi-sectoral agencies, communities, and individuals should work together to develop plans (State of the Planet, 2023).

3.4. CLIMATE CHANGE AND WATER SECURITY

Water plays a crucial role in the daily lives of humans and is vital for various economic activities in both rural and urban areas, as well as in industrial settings. Access to water can vary significantly, ranging from the convenience of a nearby tap to the necessity of walking, waiting in line, and transporting water, often over considerable distances. Water poses significant risks to human health and safety. Contaminated water can lead to severe illnesses, and floods are notorious for causing extensive damage and loss of life, making them one of the most dangerous climate-related disasters (Figure 3.4).

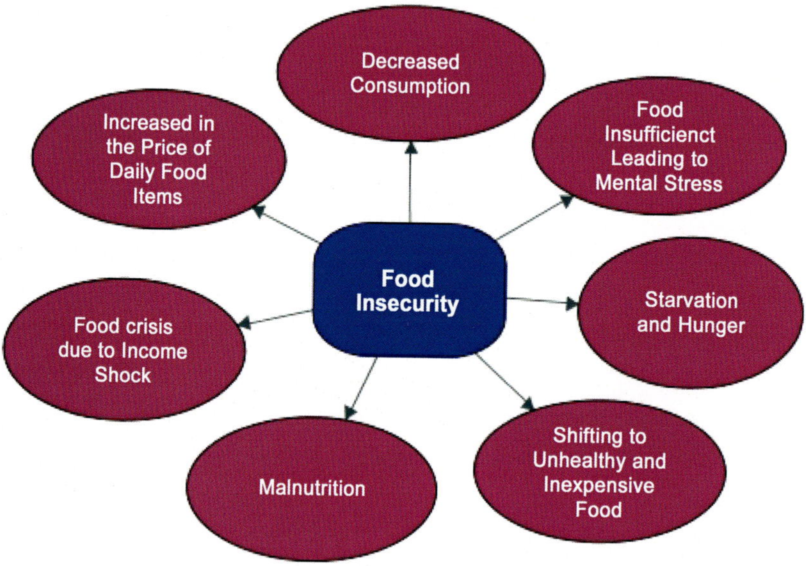

Figure 3.4. *Over 20 million people, including 10 million children, in East Africa need food and water aid.*

Source: Sohel MS, Shi G, Zaman NT, Hossain B, Halimuzzaman M, Akintunde TY, Liu H. Understanding the Food Insecurity and Coping Strategies of Indigenous Households during COVID-19 Crisis in Chittagong Hill Tracts, Bangladesh: A Qualitative Study. Foods. *2022; 11(19):3103. https://doi.org/10.3390/foods11193103.*

The global aim of ensuring that all individuals have access to water and sanitation, as well as sustainable management of this resource, is emphasized in the Sustainable Development Goal on water. Although we have come a long way, there are still 771 million people in the globe who do not have access to safe drinking water, so we still have a ways to go before we achieve this goal. All of these facets of water are included in the idea of water security.

As a result of the growing effects of climate change, such as changes in precipitation patterns, definitions usually include the need for a reliable source of safe and clean water. Also covered are problems with water supply systems and the difficulties of making sure everyone has access to water. Ensuring proper water levels for wildlife in wetland habitats is one aspect of water security, which also includes taking proper precautions against flooding and other water-related dangers. Water is also essential for maintaining healthy ecosystems.

Recognizing the profound interconnections between the social and physical systems of water resources is crucial for ensuring water security. To do so, one must be cognizant of the fact that water systems are both affected by and contribute to human activities. In order to solve the problem of water security, we must think about it from a physical and human perspective. Making progress in this area requires making sure that everyone has easy access to enough high-quality water. Ensuring accessibility of water resources requires:

- Effective provision;
- Allocation;
- Management, and pricing; and
- Social and political principles have a significant impact on these areas.

There are several interrelated challenges to ensuring and improving water security. Over 90% of all disasters are water-related, so it's important to recognize their impact. This was stressed during the 2023 United Nations Water Conference. These problems are made worse by climate change, which is especially harmful to water supplies. Consistent political and public attention is essential for making progress toward water security that is climate resilient and guaranteeing that everyone has access to safe water. We also need to implement water management plans that consider the demands of all users and increase funding (UNEP, 2024).

3.4.1. What Effects Does Climate Change Have on the Water Cycle and Security?

Climate change is a multifaceted phenomenon, encompassing gradual shifts in temperature, alterations in weather patterns, and the escalation of extreme events. The hydrological (water) cycle plays a crucial role in both environmental and societal aspects, with significant effects on water security.

The availability of surface and groundwater sources can be significantly affected by variations in precipitation patterns, such as:

- Changes in timing;
- Amount;

- Intensity;
- Form (snowfall or rainfall); and
- Alterations in atmospheric evaporative demand.

These factors play a crucial role in determining the overall water resources. The land's surface characteristics, such as land use and land cover, can undergo changes due to factors like higher temperatures and human activity. These changes, in turn, have an impact on the water cycle. The shrinkage of glaciers and snow cover in mountain environments is a result of intricate interactions between water resource systems and the melting of glacial ice. The demand for water and future water security will be influenced by various factors, including climate change (such as for irrigation) and other factors like population growth and urbanization.

It is widely acknowledged that climate change is a persistent and ongoing issue. However, there remains some uncertainty regarding the specific changes that will occur within the water cycle. The level of uncertainty surrounding future precipitation patterns is quite significant. This uncertainty arises mainly from the discrepancies observed in climate model results. Additionally, the fluctuation of precipitation over decades can lead to alternating wet and dry periods in the same area over time. In general, there will be an increase in both the frequency and intensity of floods. One can anticipate a rise in global precipitation due to the correlation between warmer temperatures and increased evaporation, which in turn allows the atmosphere to retain larger amounts of water vapor.

Water supply systems are vulnerable to prolonged droughts, as has been highlighted by recent events. The "Millennium drought" in Australia and the current drought in California are two prime examples. 'Day Zero' brought considerable attention to the fact that Cape Town's water supply was in jeopardy due to the almost empty reservoirs in the city. Although drought was not the only element at play here, the Cape Town case illustrates how climate change could jeopardize water security.

Climate change has significant implications for water security, impacting everything from the flow of transboundary river systems that span multiple territories to the availability of local unimproved water sources such as unprotected wells, springs, and surface water. The changing patterns of extremes will lead to an increase in flood risk, which will have significant consequences for the wellbeing of individuals and communities. This includes both the physical impacts of floods and the psychological effects of being exposed to risk and the subsequent recovery from trauma.

The extensive secondary impacts of climate change on food and energy production, as well as land use, have the potential to significantly disrupt water management systems (American Security Project-ASP, 2023).

3.4.2. The Management of Water Resources

The provision of water for urban uses, irrigation of crops (the single largest user of water on a global scale), and industrial operations like cooling power plants all rely on careful management of water resources. The long-term stability (or "stationarity") of our climate and water sources has been traditionally relied upon by management. However, this assumption is being challenged by climate change.

Even though water storage, demand management, and early warning systems are ways that water management systems handle some uncertainty and variability, it's unclear if these will be enough to handle the predicted speed and scale of change to water resources due to climate change.

The long-term stability of rainfall and streamflow patterns is still frequently assumed when large-scale water infrastructure projects are funded. Long operational lifetimes, however, will subject them to a far broader spectrum of variability and extremes than is now the case. This is a major performance risk, and although guidelines for making large-scale water projects more climate resilient are being developed, they are still optional and seldom used (United Nations, 2007).

3.4.3. Adjustments Required to Address the Impact of Climate Change on Water Security

Uncertainty surrounding the future of climate patterns remains a crucial factor in determining effective strategies for water sector adaptation. At present, climate models do not provide the level of spatial and temporal detail that is often necessary for effective adaptation planning. It is important to incorporate higher levels of uncertainty into water resource planning and to consider how water systems can handle various future climate conditions. This should be done in conjunction with other factors such as:

- Urbanization;
- Development; and
- Population growth.

Water resource planners and managers are adapting to these evolving challenges, as practitioners navigate shifting risk profiles and the practical implications they bring. Adaptation involves implementing technical and engineering strategies such as:

- Water storage;

- Water-saving devices; and
- Flood protection measures.

These measures are crucial for ensuring a safe water supply. Additionally, socioeconomic responses are necessary to ensure a sustainable level of demand and equitable accessibility. Early warning systems for flood or drought events can incorporate both physical improvements, such as enhanced forecasts, and social elements, like better risk communication. Engaging communities and stakeholders throughout these processes can help shape effective and resilient adaptation strategies that are customized to specific local circumstances (Wilson Center, 2014).

Considering climate change in planning is already becoming mandatory in certain jurisdictions. As an illustration, the UK's Climate Change Act mandates the publication of Water Resource Management Plans every five years, covering a minimum period of 25 years. Proposed adaptation options for the UK involve:

- Meeting formal international standards of resilience;
- Bolstering building regulations;
- Implementing water metering; and
- Improving drought management.

Although there are numerous strategies to enhance resilience and water security, there are certain challenges that need to be addressed. These challenges include disparities in access to various options and a limited amount of detailed evidence on their effectiveness, especially in the face of increasing global warming (World Economic Forum, 2023).

3.5. CLIMATE CHANGE AND ENERGY SECURITY

The vulnerability of our energy system to various climate change impacts is a cause for concern. Some of the events that we need to be concerned about are:

- Increasing temperatures and heat waves;
- Cold and snow events;
- Severe drought;
- Intense rainfall;
- Rising sea levels; and
- Hurricanes, and wildfires.

Although the effects may vary across different regions, they will persistently impact all parts of the world. In addition, the effects on one aspect of the energy system or in a specific location can have repercussions on other components of the system or different regions (American Security Project-ASP, 2023).

The energy sector can be impacted by climate change on both local and regional levels. Four significant impacts are outlined in this section.

3.5.1. Energy Supply Interruptions

Severe weather conditions and unpredictable natural calamities present substantial threats to the energy supply in different areas around the world. Both the Gulf and East Coasts are experiencing increased vulnerability to damage from flooding caused by hurricanes and rising sea levels. There is a growing concern about the potential impact of more frequent and intense precipitation events on infrastructure. In Arctic regions such as Alaska, the melting of permafrost leads to land sinking and poses a threat to fuel pipelines and other energy infrastructure.

In general, there is a trend towards warming temperatures, which can result in longer periods of drought due to the atmosphere's ability to retain more moisture. The changes in weather patterns can have significant impacts on water resources, which in turn affect the availability of energy supplies. As an illustration:

- Many power plants in the United States rely on rivers or lakes for their cooling needs;
- Water is essential for the production and refining of petroleum, natural gas, and biofuels; and
- Hydropower production in the West has been significantly impacted by water shortages.

Insufficient water supplies necessitate affected systems to seek alternative water sources or reduce their operations.

3.5.2. Disruptions to Power Distribution

Power supply to our homes and businesses is at risk due to the impact of climate change. For instance, during severe weather conditions, transmission lines are susceptible to damage such as:

- Extreme weather conditions such as snow and ice, wildfires, and strong winds have the potential to cause significant harm to above-ground powerlines and transmission towers.
- Flooding has the potential to cause significant damage to underground powerlines, as well as roads, railroads, pipelines, and storage facilities.
- In coastal areas, the destructive force of storm surge can wreak havoc on petroleum storage tanks and cause significant damage to roads and railways.

- Higher temperatures, particularly during the summer season, have the potential to impact the transmission of power.

As temperatures increase, the ability of transmission lines to carry electricity decreases. The summer months bring about increased dangers of wildfires. Wildfires have the potential to cause major disruptions to energy networks when they impact transmission towers and powerlines. In certain regions, powerlines that are faulty or have fallen (or come into contact with trees) have the potential to ignite wildfires. Due to the potential danger, certain utilities choose to deactivate powerlines in anticipation of strong winds (United Nations, 2007).

3.5.3. Pressure on the Energy System

Temperatures are increasing in all regions of the world. With the rising temperatures, there is a projected increase in energy consumption among people, particularly for cooling purposes. This increased demand may lead to a higher likelihood of blackouts or other power interruptions.

With the climate changing, it is anticipated that most people will need to reduce their energy consumption for winter heating. Nevertheless, the anticipated rise in cooling requirements during the summer is projected to surpass any potential energy savings resulting from decreased heating needs.

3.5.4. The Growing Concern Over Air Pollution and Climate Change

With the increasing demand for cooling across the world, there is a need to generate more electricity to meet this demand. Expanding energy production has the potential to result in higher emissions of specific air pollutants and greenhouse gases, which are known contributors to climate change.

3.5.5. What Can Be Done

There are several ways to mitigate the impact of climate change on the energy sector:

- **Conserve Energy:** There are numerous ways for individuals and companies to conserve energy. As an illustration, consider searching for products that are ENERGY STAR certified, like appliances and electronics. Certain utility companies even provide federal tax credits.
- **Promote Widespread Adoption of Clean Technologies:** Government and industry leaders have the opportunity to play a crucial role in promoting the expansion of renewable energy

programs, including wind and solar power, to ensure that all communities can reap the benefits. This transition will have a significant impact on reducing the emissions that are contributing to climate change.

- **Update the Infrastructure:** Utilities and government agencies can modernize energy infrastructure, addressing issues like leak-prone pipelines and aging power lines. These actions aim to:
  - Enhance resiliency;
  - Bolster safety measures; and
  - Safeguard public health.
- **Promote Fairness in Energy Distribution:** It is important for policymakers, industry leaders, and communities to work together to enhance energy affordability and promote inclusivity in energy planning. They can also ensure that the benefits from energy investment reach typically underserved communities, just like a leader would.
- **Empower Communities by Prioritizing Local Infrastructure:** Utilities, urban planners, and government agencies can all benefit from the use of microgrids. These systems and other decentralized energy infrastructure contribute to the increased resilience of electricity supplies during extreme weather events.

3.6. CLIMATE CHANGE AND HEALTH SECURITY

Many factors, both natural and man-made, affect human health and illness, and climate change is just one of them. New health threats will appear, and some old ones will get worse. Not everyone is facing the same level of risk. Consideration should be given to age, financial means, and geographical location.

Disruptions to physical, biological, and ecological systems, whether they arise domestically or internationally, can have an impact on public health within our borders. The disruptions have led to a range of consequences, including higher rates of respiratory and cardiovascular diseases, injuries, and fatalities caused by severe weather. Additionally, there have been shifts in the distribution and incidence of infectious diseases, problems with mental health, and changes in the prevalence of food and water-borne illnesses (American Security Project-ASP, 2023).

3.6.1. Climate Change and Air Pollution

One way that climate change is expected to have a detrimental effect on human health is through an increase in ground-level ozone and/or particulate matter air pollution. Ground-level ozone, a key component of smog, has been associated with various health problems such as decreased lung function, increased hospitalizations and emergency room visits due to asthma, and a rise in premature deaths (Figure 3.5).

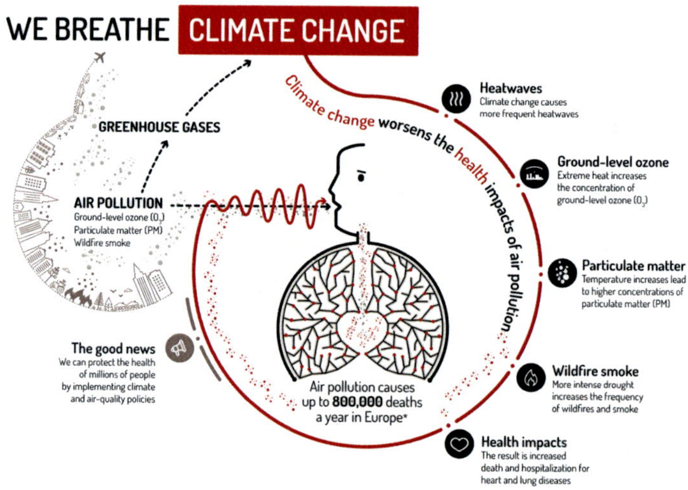

Figure 3.5. *The connection between heatwaves, air pollution, wildfires, health, and climate change.*

Source: DAHL, 2020.

Factors such as temperature, precursor chemical concentrations, and methane emissions play a significant role in the formation of ozone. Several factors, such as emissions from wildfires and periods of air stagnation, contribute to the levels of particulate matter in the air. Climate change is expected to lead to increased levels of ozone and particulate matter in specific regions, as it intensifies these different factors. Due to the deteriorating ozone and particle pollution, the rise in global temperatures could potentially lead to a higher number of premature deaths.

In the future, it is estimated that the world could see an increase in premature deaths due to the health effects of ozone and particle pollution. This projection assumes that there will be no changes in regulatory controls or population characteristics. Compared to ozone, the scientific understanding of how airborne particles react to climate change is still limited.

3.6.2. Water- and Food-Borne Diarrhea

Despite its decline in developed countries like the US, diarrheal illness remains a significant public health issue in developing nations. Diarrheal disease occurs when individuals come into contact with different pathogens found in food and water sources. Seasonal variations, extreme rainfall events, patterns of precipitation, and air and water temperatures have a significant impact on the spread of disease. Children and the elderly face a higher risk of experiencing severe consequences, particularly if they come into contact with untreated or inadequately treated groundwater. This vulnerable group will be greatly affected by such exposure.

Diarrheal illnesses like salmonellosis and campylobacteriosis tend to be more prevalent during the warmer months, although the specific patterns may differ depending on the location and pathogen. In addition, it has been found that both unusually high and low levels of precipitation are linked to a greater occurrence of diarrheal illnesses. It has also been observed that outbreaks are often preceded by sporadic increases in stream flow rates, which are commonly triggered by sudden snowmelt and changes to water treatment (American Security Project-ASP, 2023).

3.6.3. Vector-Borne Diseases

Climate plays a significant role in the spread of diseases carried by vectors such as mosquitoes, fleas, and ticks. These pathogens can cause various illnesses. Aside from climate, various factors influence the geographic and seasonal distribution of vector populations and the diseases they can transmit. These factors encompass land use, socioeconomic, and cultural aspects, pest control, access to healthcare, and human reactions to disease risk. The daily, seasonal, or annual climate fluctuations can prompt vectors and pathogens to adapt and cause shifts or expansions in their geographic ranges. These changes have the potential to impact the occurrence of disease, depending on factors such as host immunity, pathogen evolution, and vector-host interaction.

Several vector-borne illnesses, including Lyme, dengue fever, West Nile virus disease, Rocky Mountain spotted fever, plague, and tularemia, are currently a concern in globally. There are also potential threats from vector-borne infections including viruses like Rift Valley fever, chikungunya, and Chagas disease.

Shifting climate may affect the spread and occurrence of diseases carried by vectors in countries where these diseases are already prevalent. This is especially relevant due to the rise in trade and travel to tropical and subtropical areas. It's uncertain how the changing climate will impact the likelihood of people contracting diseases such as dengue fever. This is due to various factors such as vector control initiatives and lifestyle choices that aim to minimize human-insect contact, such as spending more time indoors. Several factors, such as weather patterns, changes in the environment, the presence of different animal hosts, and human behavior, can all contribute to the transmission of infectious diseases. Further extensive research is necessary to accurately determine the connections between weather conditions, the spread of disease-carrying organisms, and the occurrence of vector-borne illnesses. Additionally, it is crucial to understand the impact of shifting patterns in vectors and pathogens, as well as how these changes influence human behavior. To tackle these issues, it is necessary to improve the monitoring of human diseases and the surveillance of vectors (American Security Project-ASP, 2023).

3.6.4. Stress-Related Mental Health Disorders

Mental illness is a significant issue in the world, and extreme weather conditions can have various effects on it. There is a fascinating occurrence called "common reactions to abnormal events" which involves an increase in mental health problems following disasters.

This affects individuals who are vulnerable as well as those who have not previously experienced mental illness. These responses can vary in duration, sometimes being brief and other times lasting longer. For example, research on Hurricane Katrina survivors has shown increased levels of anxiety and PTSD. Similar results have been found after heat waves and floods. There is evidence that wildfires have significant impacts. Climate change is increasingly influencing all of these events. There have been concerns raised regarding the potential health impacts of highly stressful situations, including preterm birth, low birth weight, and complications for mothers.

Furthermore, climate change can have a greater impact on individuals with mental illness. For instance:

- Weather patterns have been observed to have an impact on suicide rates, particularly during periods of high temperatures. This correlation suggests that climate change may have a greater influence on the vulnerability of individuals with depression and other mental illnesses.
- Heat waves can have serious consequences for individuals with dementia, potentially leading to higher rates of hospitalization and mortality.
- People who are on medications that can affect body temperature regulation or even lead to overheating are more vulnerable during hot weather, particularly if they have a serious mental illness like schizophrenia.

There are other potential impacts on mental health that are not often discussed, such as the feelings of anxiety and hopelessness that can arise from the issue of climate change. Additionally, the distress caused by environmental degradation and displacement can also have a significant impact on individuals (United Nations, 2007).

3.6.5. Guaranteed Access to Nutrition

Global food production, some aspects of food quality, food prices, and food distribution networks are all expected to be threatened by climate change. Many crop yields are predicted to decline as a result of the combined effects of fluctuating rainfall, extreme weather, and increasing competition from weeds and pests for crop plants. A decline in fish and livestock production is also anticipated. Prices are predicted to rise as a result of declining food production and associated trends, such as rising petroleum prices for agricultural inputs like pesticides and fertilizer.

In comparison to developing nations, most developed countries will not be completely unaffected. A person's health can be impacted by numerous factors such as:

1. **First:** People who adhere to strict diets will face shortages of necessary foods.
2. **Second:** Food insecurity increases in tandem with food prices. In these situations, people often give in to hunger or consume foods heavy in calories but low in nutrients, which can have detrimental effects ranging from micronutrient malnutrition to obesity.
3. **Third:** It is anticipated that some foods' nutritional value will decline. Reduced plant nitrogen concentration and, subsequently, decreased protein are associated with increased atmospheric CO_2 in a number of crops, such as barley, sorghum, and soy. Crop levels of nutrients such as calcium, iron, zinc, vitamins, and sugars are predicted to be lower if soil nitrogen levels are below optimal levels. This effect can be mitigated by providing sufficient nitrogen.
4. **Fourth:** Farmers should anticipate using more herbicides and pesticides overall as a result of weeds and pests growing faster and some chemicals losing their effectiveness over time. There will be a rise in the exposure of farmers, farmworkers, and consumers to these compounds and the potentially harmful residues they leave behind.

These impacts of climate change on the nutritional content of food have a larger context. In addition to these, food distribution networks, consumer food preferences, and agricultural methods are all significant variables. Adaptation efforts can reduce the health effects of some of the challenges to food security that are anticipated (Wilson Center, 2014).

3.6.6. Severe Temperatures

For an extended period, the well-being of the public in most countries has been endangered by heat waves. The number of fatalities caused by heat waves has seen a significant increase in several cities, including Cincinnati, St. Louis, Philadelphia, Chicago, and various others. Heat stroke and its complications are among the top causes of death, along with heart disease, lung disease, and stroke. Individuals with heart, kidney, and respiratory issues have a higher probability of being admitted to the hospital during periods of extreme heat. The United States is facing an increasing number of intense summer heat waves, and climate predictions indicate that these waves will continue to grow in frequency and intensity in the future.

It seems that in recent decades, there has been a decrease in the occurrence of heat-related illness and death. This could be attributed to factors such as:

- Enhanced forecasting;
- The implementation of heat-health early warning systems; and
- Increased availability of air conditioning.

However, heat waves continue to result in preventable fatalities throughout the world. According to future projections, urban populations are expected to face greater risks to their health from heat-related impacts. This is due to:

- The presence of urban heat islands;
- An aging population;
- The ongoing process of urbanization.

One can anticipate a decrease in cold-related illnesses, injuries, and fatalities due to the milder winters resulting from a warming climate. Factors such as housing, age, and pre-existing health can influence an individual's vulnerability to winter weather. While climate change is expected to have a positive impact on reducing the number of fatalities and injuries caused by extremely cold weather, it is not expected to counterbalance the increase in fatalities caused by heat waves (American Security Project-ASP, 2023).

3.6.7. Climate Change, Allergens, and Pollen

The potential consequences of climate change include longer pollen seasons and higher pollen concentrations, which could lead to a greater number of individuals being negatively impacted by pollen and other allergens. Pollen, being an airborne allergen, can have a significant impact on our overall health and well-being. Plants, trees, grass, and even weeds have the ability to release tiny "seeds" known as pollen grains. The quantity and kind of airborne pollen are determined by the season and location. Plants can pollinate throughout the year, with a noticeable increase in activity during the spring and summer months.

Possible outcomes of climate change include:

- Changes in precipitation patterns;
- Fewer days with frost;
- Higher seasonal air temperatures; and
- Potentially increased levels of atmospheric carbon dioxide (CO_2).

These changes have the potential to impact: the duration of the pollen season; the annual length of time it lasts; the quantity of pollen produced by plants and present in the air; the effects of pollen on human health; the level of exposure to pollen; and the probability of experiencing allergy symptoms.

Exposure to pollen can lead to various allergic reactions, including symptoms of hay fever. When pollen and other allergens infiltrate the body and are mistakenly perceived as threats, the consequence is hay fever, also known as allergic rhinitis. When allergens enter the body, it can trigger a reaction in the nose, resulting in symptoms of allergic rhinitis. Signs of allergic rhinitis include:

- Sneezing;
- Nasal discharge; and
- Stuffiness.

Symptoms of allergic conjunctivitis can also be triggered by exposure to pollen. When allergens, like pollen, are inhaled into the eye, allergic conjunctivitis develops. A significant percentage of individuals who have allergic rhinitis also experience allergic conjunctivitis, with the prevalence increasing to approximately 30% in the general population. When experiencing allergic conjunctivitis, one may observe:

- Redness in the eyes;
- Excessive tearing; and
- Itching in the eyes.

Individuals with respiratory conditions, such as asthma, may be more susceptible to the effects of pollen. There is compelling evidence indicating that exposure to pollen can potentially trigger asthma attacks and result in an increase in hospital admissions for respiratory illnesses. Each year, pollen results in over $3 billion in medical expenses, with prescription medicine making up almost half of that amount. If the pollen concentrations are higher and the pollen seasons are longer, you might notice an increase in sensitivity to allergens. Individuals with asthma may encounter a decline in their symptoms, leading to decreased productivity at work or school (UNEP, 2024).

Extreme precipitation and rising temperatures can worsen indoor air quality problems. For instance, they have the potential to encourage the growth of mold indoors, which can worsen asthma symptoms and make it more challenging for individuals to manage their asthma (American Security Project-ASP, 2023).

3.6.8. Heavy Rain, Flooding, and Droughts

Heavy rains are increasingly occurring throughout the world and are projected to continue affecting most places globally. Severe flooding has become increasingly frequent in certain regions, attributed to rising levels of both overall precipitation and intense rainfall events. For instance, every year, almost 98 lives are lost in the United States due to flooding, making it the most fatal weather-related disaster. Tropical storms and flash floods are responsible for the highest number of casualties.

Once a storm has subsided, there may be additional hazards that arise, in addition to the immediate health concerns associated with flooding and intense rainfall. While it is important to consider other factors that may contribute to these correlations, there have been reports indicating a rise in waterborne disease outbreaks following periods of heavy rainfall. Indoor air quality problems may arise in the future due to mold contamination caused by water intrusion. Buildings become more susceptible to water seepage after being damaged by a hurricane. Populations residing in damp indoor environments are more prone to experiencing asthma, coughing, wheezing, and other upper respiratory tract symptoms. Additionally, they are at a higher risk of developing pneumonia, respiratory syncytial virus (RSV), and RSV pneumonia (United Nations, 2007).

Drought poses a significant threat to public health and safety, as it represents the opposite extreme of precipitation. During a drought, a variety of environmental health risks can intensify. These include:

- Wildfires;
- Dust storms;
- Scorching temperatures;
- Sudden floods;
- Reduced water supply; and
- Deterioration in water quality.

For instance, in Arizona and California, the occurrence of dust storms due to drought has been associated with a rise in coccidioidomycosis (valley fever), a fungal disease, and a deterioration in air quality caused by particulate matter.

3.6.9. Wildfires

Wildfires are increasingly prevalent in certain regions of the world and pose a growing threat to numerous forests as a result of climate change. Extended periods of high temperatures can worsen dry conditions and potentially lead to wildfires in certain regions. Wildfire smoke contains a range of harmful substances, such as:

- Particulate matter;
- Carbon monoxide;
- Nitrogen oxides; and
- Volatile organic compounds.

These pollutants can greatly reduce air quality, not only in the nearby areas but also in downstream regions.

Smoking has been found to significantly increase the risk of chronic obstructive pulmonary disease (COPD), respiratory infections, asthma, bronchitis, chest pain, emergency department visits, and hospitalizations for cardiovascular and respiratory illnesses. The global health risks associated with smoke from landscape fires have been found to be responsible for a significant number of annual fatalities. Increasing temperatures are anticipated to amplify the risk of wildfires and the resulting emissions, which can adversely impact human health (World Economic Forum, 2023).

3.7. CLIMATE CHANGE AND DEFENSE STRATEGIES

It is crucial for future defense strategies to consider the impact of climate change, as it has the potential to significantly amplify threats faced by the defense forces.

3.7.1. Adapt

Due to climate change, future military forces may encounter varying deployment patterns, operational environments, and mission types. Mass migration from inhospitable regions and crop failures that disrupt food supplies are examples of second-order effects of climate change that require urgent attention and action. Resettlement planning for refugees and displaced people is acknowledged in the United States through Executive Order 14013.

The damage caused by Hurricane Michael in 2018 to Tyndall Air Force Base in Florida amounted to approximately $4 billion. This event serves as a clear example of how climate change can pose a direct threat to defense forces and their facilities. The efficiency of air lift is reduced in hot, less dense air, and similar capabilities can be compromised by a warmer climate. It is crucial for armies to have supply lines that are capable of sustaining themselves on the battlefield, even in challenging weather conditions.

3.7.2. Mitigate

The production of greenhouse gases by the defense sector has significant implications for the environment. As an example, the UK Ministry of Defense (MOD) is responsible for a significant portion of the federal government's greenhouse gas emissions, water consumption, and trash production. The defense industry's carbon footprint is significantly impacted by major defense manufacturers, both domestically and internationally. This is due to their contribution to pollution, resource consumption, and indirect emissions (United Nations, 2007).

Defense encompasses industries such as aviation and maritime, which encounter some of the most challenging decarbonization issues. Additionally, it involves energy-inefficient estates that have aged over time, posing significant obstacles to mitigating these effects.

Inaction will not be tolerated. If decarbonization does not occur, defense budgets may be at risk of being impacted by carbon taxes and remaining reliant on fossil fuels. This could potentially lead to increased costs as the world transitions to alternative energy sources. This could potentially hinder the investment in critical capabilities.

Furthermore, neglecting the potential of modern technology and disregarding the benefits of transitioning to renewable energy poses significant risks. The defense sector risks losing its credibility and impact if it fails to take action against climate change. Combat units equipped with distributed energy generation, water-and energy-efficient assets, and reduced logistical support are more resilient in the face of fuel and supply shortages. Electric vehicles have made significant strides in terms of performance, reliability, and affordability when compared to their traditionally powered counterparts. From a financial and strategic standpoint, militaries can greatly benefit from adopting innovative and environmentally-friendly technology (Wilson Center, 2014).

3.7.3. Putting a Climate Change Response at the Top of the Agenda

The importance of addressing the challenge of climate change and its impact on military operations is increasingly acknowledged.

It is crucial to prioritize actions based on their impact, taking into account the scale of the task and the various demands of defense activity. Defense agencies can enhance their preparedness for climate change by conducting comprehensive assessments of the physical and transitional risks they may encounter, considering the potential consequences of different future climatic scenarios. Despite its complexity, the private sector is rapidly embracing this approach, which can be efficiently implemented with the assistance of predictive analytics.

These risks and opportunities may not always receive the attention they deserve, given the various defense interests at play. However, it is crucial that we actively participate in the climate change debate.

Defense organizations must prioritize the concerns of allies, governments, and citizens in a world where extreme weather events and rising sea levels are increasingly prevalent. It is crucial to stay ahead of, or even surpass, the response of adversaries. By taking the right actions, we can achieve a more effective defense capability and make progress in reducing the rate of global warming (World Economic Forum, 2023).

CHAPTER 4

REGIONAL PERSPECTIVES ON CLIMATE SECURITY

Contents

INTRODUCTION

Emerging humanitarian crises are being exacerbated by the effects of climate change, compounding the challenges faced by already vulnerable communities grappling with poverty and conflict. All of these regions have one thing in common: they are highly vulnerable to the impacts of climate change, despite relatively low greenhouse gas emissions in some regions.

4.1. CLIMATE SECURITY IN NORTH AMERICA

The North American populations and the systems they depend on are facing significant threats due to the escalating risks linked to climate change. These risks are growing at an alarming rate and pose a serious danger to the natural, managed, and human systems in the region.

Even with a confidence level of 1.5°C, the safety of humans in North America, especially in coastal regions like Mexico, the Hawaiian Islands, the Gulf of Mexico, Atlantic Canada, and the south-eastern United States, will be at risk due to the increasing sea levels and more frequent and severe storms and hurricanes (US EPA, 2024).

4.1.1. Effects on the Ecosystem

It is quite evident that the changing ecosystems due to increasing temperatures in the air, water, ocean, and ground have had a significant impact on the redistribution and mortality of fish, bird, and mammal species.

The changes occurring in Arctic ecosystems due to climate change are undeniably significant. It is highly likely that the impacts of climate change on various ecosystems will significantly alter ecological processes, and that other threats to endangered and iconic species and their habitats will be intensified (State of the Planet, 2023).

4.1.2. Effects on Health

According to the United States Department of State (2023), climate change has had a negative impact on the well-being of people in North America. The risk of death and illness significantly rises during periods of extreme heat, affecting individuals of various ages, genders, locations, and socioeconomic statuses. It is highly likely that health risks will increase in the coming century, regardless of the emissions scenarios being considered. However, the severity and magnitude of these impacts depend on the implementation of adaptation strategies. It is anticipated that there will be an increase in both heat-related deaths and heat-related illnesses due to the effects of global warming (US EPA, 2024).

4.1.3. Effects on Economic Processes

Throughout North America, the infrastructure of supply chains and trade has been disrupted, leading to negative impacts on economic activities due to climate hazards and extreme events. The potential damages caused by climate change, whether in the market or non-market sectors, are expected to be significant.

4.1.4. Effects on Food

Climate-induced changes in North American food production have raised concerns about future food and nutritional security. The suitability of farming and fishing in North America will be greatly affected by climate change, leading to significant changes in the ranges of crops, livestock, fisheries, and aquaculture products. Unfortunately, the impact on the production of these goods is expected to worsen (State of the Planet, 2023).

4.1.5. Effects on Water

The deteriorating state of freshwater management infrastructure and excessive utilization of limited water resources have intensified the risks, especially in the western United States and northern Mexico. During periods of drought and earlier snowmelt runoff, there will be economic losses and increased demands on limited groundwater.

This will worsen water scarcity, especially during the summer peak water demand period in regions with extensive irrigated agriculture (United States Department of State, 2023).

4.1.6. Effects on Urban Areas and Communities

The frequency and severity of climate hazards and extreme events in North American cities and settlements have worsened infrastructure damage, livelihood losses, heritage resource damage, and safety concerns.

Indigenous peoples are particularly impacted by these effects as a strong and sustainable environment is deeply intertwined with their culture, identity, economy, health, and overall well-being. Urban, riverine, and coastal flooding presents a substantial danger to coastal ecosystems and communities, potentially disrupting transportation and trade infrastructure and compelling individuals to evacuate their residences (Figure 4.1). The detrimental effects of massive wildfires extend beyond the destruction of lives, livelihoods, property, and essential infrastructure. They also contribute to air pollution and water contamination in urban areas, posing a significant threat to both human health and economic activity (US EPA, 2024).

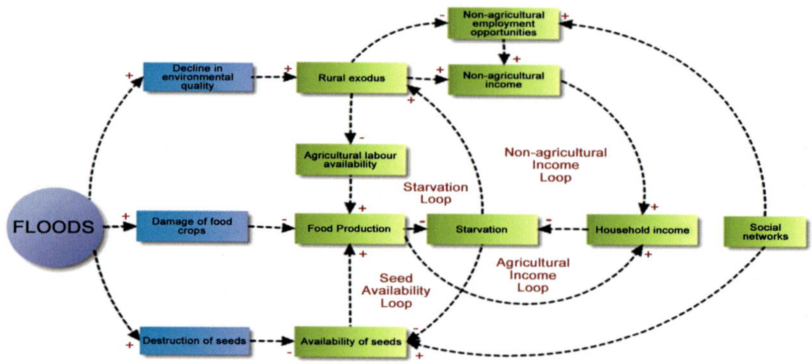

Figure 4.1. *The impact of floods.*

Source: Armah FA, Yawson DO, Yengoh GT, Odoi JO, Afrifa EKA. Impact of Floods on Livelihoods and Vulnerability of Natural Resource Dependent Communities in Northern Ghana. Water. *2010; 2(2):120-139. https://doi.org/10.3390/w2020120.*

4.1.7. Obstacles in Addressing Climate Change

The division in North American public and policy spheres has been caused by the spread of false information and the politicization of climate change science. Unfortunately, this has hindered progress in addressing climate change, despite the overwhelming scientific consensus on the human impact on our climate. There are certain groups that have been spreading inaccurate information and using misleading language to undermine the credibility of climate science and minimize the gravity of the issue.

Adaptation planning and implementation are unfortunately being delayed due to the current public polarization and misunderstanding of climate risks. Although community-level planning is crucial for making adaptation responses and disaster preparedness more context-specific, it can sometimes face challenges due to policy misalignment at various levels of government. Ensuring that policies and infrastructure are resilient to climate change demands:

- Meticulous planning;
- Widespread support; and
- Effective coordination to secure sufficient funding.

4.1.8. Possible Courses of Action

Implementing climate impact predictions in decision-making processes, both in the short and long term can help reduce future risks in a fair, inclusive, and collaborative way. Efforts to modify land-use impacts, consumption patterns, and economic activities, and emphasize nature-based solutions need to be coordinated and integrated through fair and inclusive policy, as current practices alone will not suffice (State of the Planet, 2023).

To effectively address climate change, it is crucial to prioritize indigenous self-determination, recognize the rights of Indigenous peoples, and encourage adaptation strategies rooted in Indigenous knowledge. Effective and equitable climate solutions can be achieved through:

- Comprehensive planning;
- Implementation, and coordination across sectors; and
- Jurisdictions support.

4.1.9. Building a Future Resilient to Climate Challenges

The projected escalation of critical threats to North America is a pressing concern that demands immediate action to keep global warming below 1.5°C. In order to ensure the well-being of North Americans, promote sustainable livelihoods, and protect long-term biodiversity and economic productivity, adaptation measures with a strong focus on fairness must be promptly and comprehensively put into action. With the implementation of strong mitigation measures, transformative, long-term adaptation action can effectively tackle the rapidly intensifying impacts by minimizing risk and enhancing resilience (US EPA, 2024).

4.2. CLIMATE SECURITY IN EUROPE

Given the intricate threats presented by climate change, climate security has emerged as a paramount concern in Europe. According to research conducted by the European Union, Europe could potentially see a three-degree increase in average temperature by the end of this century if significant reductions in greenhouse gas production are not achieved globally (Division and Nikolayeva, 2017).

Water scarcity is a pressing issue that affects a significant portion of the European population. The continent's infrastructure would suffer a significant economic setback. The impact of drought on farmland would increase sevenfold. There might be areas where agricultural yields experience a decline of 20% or even more. If current trends persist, the number of heat-related deaths could rise by 133,000 annually (Lazard and Youngs, 2021).

In addition, Europe is susceptible to the effects of climate change on both a local and global scale. This vulnerability is a result of its wide-ranging international political, commercial, cultural, and financial connections (Zwolski and Kaunert, 2011). The interconnectedness of supply chains, finance, and the current pandemic has highlighted the fragility of our interconnected world and emphasized the importance of global cooperation in addressing global challenges (Lazard and Youngs, 2021; Figure 4.2).

Figure 4.2. *Melting Arctic ice uncovers valuable untapped mineral and oil deposits.*

Source: By NASA Goddard Space Flight Center from Greenbelt, MD, USA – Ponds on the OceanUploaded by PDTillman, Public Domain, https://commons.wikimedia.org/w/index.php?curid=15837631.

Meanwhile, there is a concern that development gains globally may be reversed due to a combination of factors in politically unstable regions. This could potentially lead to an increase in the number of crises that Europe will have to address (Youngs, 2021). There have been significant shifts in global power dynamics, along with a growing demand for assistance through humanitarian and development initiatives. Additionally, there has been a noticeable surge in the number of individuals displaced due to climate-related catastrophes like floods, droughts, desertification, and storms. In light of the potential displacement of numerous individuals as a result of climate change, it is projected by experts that asylum applications to the European Union could experience a significant increase of 28% by the conclusion of this century, even under the assumption of moderate emissions (Lazard and Youngs, 2021).

4.2.1. Europe's Growing Climate Security Concern

When new information about the effects of climate change on national security emerged, European nations and institutions acted promptly to issue political declarations. The European Security Strategy from 2003 was the first document to recognize climate change as a potential threat. In 2007, the UK took the lead in organizing the inaugural UN Security Council discussion on climate security. In 2008, High Representative Javier Solana released a report that further heightened awareness. Based on this information, it is important to recognize that climate change can contribute to the escalation of existing security concerns, including issues like food and water shortages, pandemics, and population displacement (Lazard and Youngs, 2021). It is worth mentioning that the Solana paper argued that taking action would help address global fragility, which directly impacts Europe's security. Since then, climate security has gained significant attention in Europe, thanks to a series of influential reports, events, and remarks.

The Green Diplomacy Network, established in 2003, was merged with the European External Action Service (EEAS) in 2012 with the aim of enhancing the incorporation of environmental considerations into EU foreign policy **(Lazard and Youngs,** 2021). As stated in the European Union's 2016 Global Strategy on Foreign and Security Policy, climate change and environmental degradation have the potential to worsen conflicts and act as a catalyst for water and food scarcity, pandemics, and displacement. An essential aspect of this strategy revolves around climate security. The "Conclusions on Climate Diplomacy" were released in February 2018 by the Foreign Affairs Council, which is a gathering of EU Ministers of Foreign Affairs. The report acknowledged that climate change has significant implications for international security and stability. During their 2018 meeting, the Council made a decision to enhance the integration of climate change and security in various areas such as:

- Policy dialogue;
- Conflict prevention;
- Development and humanitarian action; and
- Disaster risk strategies.

The conclusions of the 2019 and 2020 Foreign Affairs Council (Council of the EU 2019; 2020) express similar sentiments.

While there has been an increase in the importance placed on climate security in European foreign policy, it remains challenging to identify concrete and bold actions taken to address the security threats posed by climate change.

The European Union (EU) possesses a vast array of resources, such as its influential role in development and global climate policy, along with various tools and instruments at its disposal to tackle the impact of climate change on international security, as highlighted by Javier Solana in 2008. Furthermore, Europe's expertise in conflict prevention, crisis management, and post-conflict reconstruction, along with its commitment to effective multilateralism, make it a strong contender for addressing the security challenge. Based on the report, it is suggested that the European Union should take proactive measures to prevent conflicts and be prepared to swiftly respond when they arise. The country should also play a prominent role in global discussions and collaborate more extensively with other nations to:

- Address climate change;
- Enhance governance; and
- Sustainably manage natural resources.

After a decade, on June 22, 2018, Federica Mogherini, the successor to the High Representative, took charge and organized an important meeting called Climate, Peace, and Security: The Time for Action. The goal was to showcase Europe, including the EU, as a significant contributor to global efforts on climate change and international collaboration for peace and security. Attendees at the meeting stressed the importance of the EU enhancing its capabilities to better tackle the security consequences of climate change (Lazard and Youngs, 2021). The event resulted in the following recommendations:

4.2.2.1. Strengthening the Link Between Climate and Security

The June 2018 meeting emphasized the need for greater attention to the climate-security connection in discussions at the highest political level, both nationally and internationally. European countries and institutions have consistently prioritized climate security with determination and conviction. In early 2019, the EU Foreign Affairs Council (FAC) restated its conclusion that climate change worsens threats and expressed their worry about the vulnerability of countries that are least equipped to tackle climate change (Lazard and Youngs, 2021).

In November 2019, the European Parliament acknowledged the presence of a climate emergency, a decision that was influenced by activists like Greta Thunberg. In January 2020, European Union Foreign Ministers gathered once again to discuss strategies for enhancing climate diplomacy. They emphasized their commitment to making climate action a key focus of the EU's external policy. The Council emphasized the importance of cooperation between the High Representative, Commission, and member states in developing effective strategies for promoting climate diplomacy.

The European Union Member States and the United Kingdom have consistently highlighted the significance of climate security in discussions at the UN Security Council. Sweden, Italy, the Netherlands, Belgium, and Germany have all placed great importance on climate security throughout their time on the Council. The United Kingdom and France, as two prominent European nations with permanent membership on the Security Council, have consistently supported the issue, starting with the debate hosted by the United Kingdom in 2007. Following that, there have been numerous discussions. After the high-level event in June 2018, Germany and Nauru came together to form the Group of Friends of Climate and Security. This group provides a platform for New York-based delegations who are interested in taking additional measures to address climate security. It serves as an informal network where members can connect and collaborate. The group started with 25 founding countries, but it has since grown to encompass more than 50 members. Among these members, around 20 are European nations, establishing it as the largest regional grouping by a considerable margin. On an international level, European countries have been actively advocating for greater global attention to climate security.

Climate security has become increasingly important in Europe's political negotiations with other regional organizations. In 2018, the EU and the African Union (AU) came to an agreement called a Memorandum of Understanding (MoU) regarding Peace, Security, and Governance. The agreement required both entities to work together on climate-related security threats in the areas of peace and security policy. The objective was to strengthen the capacity to tackle the challenges posed by the interaction of climate change with social, economic, demographic, and political factors, to mitigate instability, insecurity, and conflict. Without a doubt, these efforts have broadened the political landscape for discussing the security implications of climate change. Climate security has become a prominent topic on the Security Council's agenda, even though some Permanent Members were initially skeptical (Lazard and Youngs, 2021).

Notably, these changes have led to adjustments in the Security Council's resolutions concerning specific countries or regions. The Council's resolution (#2349) in March 2017 recognized the negative effects of climate change and ecological changes, among other factors, on the stability of the Lake Chad subregion. These effects encompass a range of challenges, such as:

- Water scarcity;
- Drought;
- Desertification;
- Land degradation; and
- Food insecurity.

In addition, it highlighted the significance of governments and the United Nations carrying out comprehensive risk assessments and implementing efficient risk management strategies regarding these factors. Since that time, there have been around 12 more resolutions of a similar nature. The majority of these resolutions have had a geographical focus, requiring multiple UN peace operations to include climate security in their planning and monitoring (Division and Nikolayeva, 2017).

Resolution 2423, adopted in June 2018, recognizes the significant consequences of climate change on the stability of Mali. MINUSMA (United Nations Multidimensional Integrated Stabilization Mission in Mali) and the Malian government should take into account these ecological changes, among other factors, when planning and implementing their activities, programs, and strategies.

Germany, currently holding the rotating Presidency of the Security Council in July 2020, put forward a comprehensive resolution on climate security. This resolution draws parallels with Resolution 1325 (2000), which specifically addresses Women, Peace, and Security. In June 2019, a coalition of NGOs and thinktanks, led by Adelphi and supported by the German Federal Foreign Ministry, released the "Berlin Call for Action: Improving the Climate for Peace." This document emphasizes the importance of prioritizing risk-informed planning, strengthening capacity for action, and improving operational responses as crucial foreign policy objectives. A proposed institutional innovation in the resolution involves creating a Special Envoy for Climate Security. This envoy would have a direct reporting line to the UN Secretary-General and would work towards mobilizing greater international efforts to tackle climate security (Lazard and Youngs, 2021).

Certain permanent members of the Security Council remain apprehensive about the real or perceived broadening of the Council's authority to encompass environmental matters. So far, this has resulted in the use of general language when it comes to climate security and a hesitancy to extensively address climate security in resolutions that focus on particular regions. However, it's hard to imagine that climate security would have gained global recognition without the unwavering determination and perseverance of European nations and institutions, much like a president leading the charge.

4.2.2.2. Give Top Priority to Implementing Paris

Another area that requires attention is the need to employ strong political and diplomatic efforts to enhance the implementation of the 2015 Paris Agreement. The Paris Agreement plays a crucial role in preventing climate tipping points, such as:

- The potential loss of the Greenland ice sheet;
- The degradation of the Amazon rainforest; and
- Disturbances to the Gulf Stream.

Tipping points can lead to significant climate-related security risks. The Paris Agreement marked a significant milestone in European climate activism, particularly in light of the challenging Copenhagen meeting in 2009, which concluded with a tense atmosphere. The agreement was made possible through the dedicated efforts of multiple European diplomats who engaged in extensive shuttle diplomacy over the course of 18 months leading up to the 21st Conference of the Parties (COP) in December 2015. Foreign ministers, such as Laurent Fabius, the former French Foreign Minister and COP co-President, highlighted the interconnections between climate change and security to emphasize the importance of a comprehensive agreement (Lazard and Youngs, 2021).

Following this, European institutions have been working towards strengthening their national climate goals and promoting the broader implementation of the Paris agreement worldwide. The authors highlight their concerns about the US's decision to withdraw from the Paris agreement and the lack of substantial progress made by other G20 countries in reducing global emissions.

In 2020, the European Union had plans to put forth a revised and more ambitious target for its 2030 Nationally Determined Contributions (NDCs). At the same time, the European Union and European nations have been actively supporting other countries in enhancing their own Nationally Determined Contributions (NDCs). In December 2019, the European Council gave its approval to the European Green Deal, as a direct response to the EU Parliament's declaration of a climate emergency the previous month. The main goal of these policy initiatives is to attain climate neutrality in Europe by 2050. In the same month, Ursula von der Leyen, the President of the European Commission, appointed Frans Timmermans as the Executive Vice President of the European Commission in charge of supervising the European Green Deal. On January 15, 2020, the European Parliament voted in favor of endorsing the deal and expressed the need for a higher level of ambition.

According to recent research, the EU Budget for the period 2021–2027 will require the EU to allocate 25% of its budget towards climate action. This represents a significant increase from the previous period's allocation of 20%.

The European Union has set more ambitious targets for reducing greenhouse gas emissions in the near future. They aim to achieve a reduction of at least 50%, and potentially up to 55%, compared to 1990 levels by the year 2030. Climate change poses a formidable challenge that requires collective action. Europe's contribution to global emissions stands at a modest 9%, and it is projected to decline in comparison to other nations (Division and Nikolayeva, 2017).

The security of Europe relies heavily on the need for other nations, including major emitters such as the US, China, and India, to make significant changes to their energy consumption and greenhouse gas emissions. Europe, as a significant player in trade and finance, can utilize various strategies to foster climate action in other regions and drive widespread innovation. The European

Green Deal includes provisions for implementing carbon tariffs on countries that do not adequately decrease their greenhouse gas emissions within a designated period. However, it is important to note that the influence of Europe on the reduction of greenhouse gas emissions in other countries is directly related to the size of a country's emissions. This influence is achieved through various means such as trade, development, and foreign policies. Typically, Europe has limited influence over the primary polluters when compared to the smaller, economically disadvantaged nations that benefit from its development assistance. Small countries do not have a substantial impact on global warming (Division and Nikolayeva, 2017).

4.2.2.3. Optimize Early Warning Systems to Maximize Effectiveness

It is crucial for Europe to improve its methods of evaluating and predicting climate risks. By implementing this system, European policymakers would gain the ability to swiftly identify climate-related hazards such as droughts, storms, and floods. These hazards have far-reaching consequences on social and economic stability, not just within Europe but also on a global scale (Zwolski and Kaunert, 2011).

European nations have shown considerable global interest in this region. From 2013 onwards, the European Commission has been actively involved in providing crucial support to the INFORM index, a global initiative carried out in partnership with the Interagency Standing Committee. This initiative aims to provide objective and concise information about the probability of humanitarian crises.

In 2017–2018, Sweden played a significant role in the Security Council by actively supporting and providing financial contributions to the establishment of the Climate Security Mechanism (CSM). The CSM was launched in late 2018 and is a collaborative effort led by the UN Environment Programme (UNEP), the UN Development Programme (UNDP), and the UN Department of Political and Peacebuilding Affairs (DPPA). The main objective of the CSM is to function as a central hub for collecting information from different UN entities. The primary purpose of this initiative is to provide timely and precise guidance to the leadership of the United Nations and the Security Council regarding the growing security threats associated with climate change.

To improve collaborations on early warning systems, it is important to strengthen partnerships with regional entities such as the African Union and the European Union. From the very beginning, the CSM has been diligently working to create a community of practice within the UN system. This community aims to seamlessly incorporate climate risks into the operations of more than

25 agencies, funds, and programs. The organization has developed a unique method for assessing climate-related risks and is now initiating the deployment of experts to local programs. However, there is some uncertainty regarding the ability of the CSM to provide clear guidance on politically contentious issues. The tripartite arrangement does not have a central governing body and requires intricate approval processes for each decision. Consequently, the mechanism maintains a natural sense of caution and is shaped by the inertia of institutions.

There have been additional measures taken to improve early warning systems at the European level. In 2014, the EEAS implemented its own early warning system, which is currently under the supervision of the Division for the Integrated Approach for Security and Peace (ISP) (Lazard and Youngs, 2021). The main goal of this early warning system is to ensure that timely information is provided to both military and civilian crisis management operations regarding peace and security activities. This includes a broad spectrum of activities, such as:

- Analysis;
- Initiatives to prevent conflicts;
- Strategies for mediation;
- Plans to prioritize peacebuilding; and
- The deployment of multilateral peace operations.

Every year, the ISP utilizes a mix of qualitative and quantitative data to identify a group of countries that may be prone to conflict in the next four years.

The EU's early warning system has incorporated a climate aspect into its assessment of which situations should be given priority for conflict prevention. This involves an analysis of the vulnerability to climate change provided by DG CLIMA. The European Union now includes a dedicated section on climate change and catastrophes in its assessment of the inherent risks of conflict.

Climate change is taken into account when formulating proactive measures. The Global Conflict Risk Index (GCRI) is a quantitative measure offered by the European Commission's Joint Research Centre to assist in the implementation of an early warning system. The index compiles publicly available data on 24 different factors and utilizes statistical regression models to assess the probability and intensity of violent conflict. In 2018, the GCRI began including a climate-related variable, specifically data on water stress (Lazard and Youngs, 2021). However, it is limited to only one dimension.

The GCRI does not incorporate data regarding natural disasters or other climate shocks, such as those caused by extreme weather events. There are also concerns about the effectiveness of this influence on risk analysis and action in the field. It is worth noting that around two-thirds of EEAS personnel are assigned to geographical desks. At this particular location, the EEAS undertakes significant and demanding tasks. As a result, it may not always be immediately apparent how much these broad thematic issues impact the day-to-day planning of activities. Nevertheless, the system's elegant concept can sometimes be disrupted by the reality when Member States have differing views on the best course of action to take early on. This situation could potentially hinder the European Union's ability to prevent a crisis from escalating (Zwolski and Kaunert, 2011). In addition, several European countries are now supporting their own individual early warning systems, which may also mirror the ones at the regional level. The Netherlands is actively supporting the Water, Peace, and Security Partnership, which includes a strong early warning component. Another example is the creation of a new Risk Informed Early Action Partnership, which was established in October 2019 and includes various partners. The funding for this initiative will be provided by the UK government until COP 26. The project's objective is to ensure that a staggering number of one billion individuals are covered by new or improved early warning systems (Zwolski and Kaunert, 2011).

There seems to be a need for more analysis on how the different components of the early warning systems ecosystem interact and contribute to policy impact. Additionally, it would be beneficial to focus on minimizing duplication within this ecosystem. European governments and EU research funds offer significant support to a vast network of researchers and thinktanks across Europe who are committed to studying climate security. This group comprises a wide range of knowledgeable individuals who diligently produce reports and coordinate a multitude of events aimed at increasing awareness, disseminating information, and delivering training. This method of early warning and conflict analysis operates in a decentralized manner. This resource provides significant insights and data on the potential emergence of new climate-related security risks, including their location, timing, and manner of occurrence. This group has been able to thrive and grow thanks to the consistent backing it has received from European governments and institutions over the years. To enhance the effectiveness of this community, it is advisable to implement a more structured framework and leverage collective analysis to expedite on-the-ground actions in a more efficient manner (Lazard and Youngs, 2021).

4.2.2.4. The Need to Emphasize Resilience and Prevention

Another important aspect to consider is the emphasis on prevention and resilience in both domestic and international policies. According to the EU, prevention should be focused on:

- Promoting sustainable livelihoods;

- Ensuring fair use of natural resources; and
- Implementing early warning systems.

Achieving this goal may prove to be quite challenging, as it requires more than just mere words to accomplish.

Before 2018, the focus was on the importance of resilience and prevention. According to the 2016 EU Foreign and Security Policy Global Strategy, it has been widely recognized that it is more efficient and effective to prevent conflicts rather than deal with them after they have already erupted. Conflicts tend to become increasingly difficult to resolve as time goes on. The EU emphasized the need to intensify efforts in prevention and monitoring, focusing on root causes such as human rights violations, inequality, resource stress, and climate change.

According to the European Green Deal, the European Union aims to collaborate with all partners in order to enhance climate and environmental resilience. This is done to prevent:

- Potential conflicts;
- Food insecurity;
- Population displacement; and
- Forced migration.

Additionally, the EU is committed to supporting a fair and equitable transition on a global scale (Division and Nikolayeva, 2017).

A significant portion of the European funds allocated to climate change adaptation, both domestically and internationally, is directed toward enhancing "resilience." Since 2008, the Global Climate Change Alliance Plus (GCCA+) has served as a support channel for policy dialogue and targeted climate action in developing countries, with the backing of the EU. The Global Alliance for Resilience Initiative is funded by GCCA+. This initiative was launched in 2012 with the aim of enhancing food security and resilience in nine Sahel countries. The "Great Green Wall" initiative, which spans 8,000 kilometers across the Sahel region, aims to reforest and restore degraded land with the support of the European Union. This African-led project is a multi-billion euro endeavor that seeks to address environmental challenges in the region.

The focus of EU humanitarian spending is on resilience. According to ECHO's report in 2016, a significant 43% of projects funded by the EU after the introduction of the "resilience marker" in 2014 incorporated disaster risk reduction activities. The IcSP and the EUTF play a crucial role in addressing climate insecurity by focusing on building resilience. For instance: The South Sudan Rural Development Strengthening Smallholders' Resilience (SORUDEV SSR) project has been granted a substantial funding of €15 million from the

EUTF. The aim of this funding is to address the pressing issues of conflict and extreme climate vulnerability in the region.

Resilience is a key focus for many European donors in their development and conflict prevention programs. The BRACED initiative, funded by the UK, aims to incorporate disaster risk reduction and climate adaptation into development strategies in the African Sahel and South and Southeast Asia.

Assessing the level of resilience that Europe has developed in climate-vulnerable areas is quite challenging. This is partly because resilience can be quite elusive. The term is used differently by the foreign policy, humanitarian, development, and defense communities.

The concept of resilience is highly relative. You must consider what it is being compared to and what you are referring to when you mention resilience. It challenging to monitor progress or demonstrate success. It becomes impossible to determine what could have occurred without the resilience programs. Therefore, the concept of "resilience" has been widely used and applied to various activities, although it has become somewhat of a buzzword. Therefore, it may not have a significant impact on promoting the necessary alignment between purpose and action in European foreign policy to effectively tackle climate insecurity (Lazard and Youngs, 2021).

4.2.2.5. Addressing the Potential Security Implications of Climate Policies

On the issue of climate change, Europe considers itself and positions itself as a world leader. If there is one area where the world needs our leadership, it is protecting our climate. For the sake of our economy, our people, and the earth, the European Green Deal is an absolute necessity. Investing in a green economic transition through promoting energy efficiency and renewable energy production are two primary goals of the European Union's climate strategy. The second goal is to assist countries in adapting to the climate change that has already been "locked in" due to previous emissions.

December 2019 saw the unveiling of the Green Deal for Europe by the Commission. By 2050, the European Union (EU) hopes to have achieved its lofty goal of becoming a resource-efficient, competitive economy that produces no net greenhouse gas emissions. Billions of euros will be allocated by the EU to achieve these objectives.

According to Zwolski and Kaunert, (2011), the European Union allocated a fifth of its budget from 2014–2020 to climate protection. The European Commission has stated its intention to raise this allocation to 25% of its budget for the years 2021–2027 in an endeavor to incorporate climate action into all EU programs. In 2018, developing nations received €21.7 billion in public climate funding from the European Union, its Member States, and the European Investment Bank—making them the largest global contributor (Lazard and Youngs, 2021). Countries all over the globe will need to adapt to climate change, and this type of effort could help them do it before the planet reaches the hazardous tipping point that would cause runaway climate change. Countless security threats could be averted if this is accomplished.

According to Zwolski and Kaunert, (2011), it may also bring about strategic advantages for Europe. For instance, switching from imported fossil fuels—which Europe doesn't produce much—to renewable energy generated at home would lessen dependence on Russian gas, ease Arctic resource competition, shift Gulf geopolitics, and change the strategic importance of the straits of Hormuz (Zwolski and Kaunert, 2011). While Europe must act swiftly to address the security threats presented by climate change, the continent must also be cognizant of the potential unforeseen security consequences of its own climate policies. Rapid and widespread changes to energy production are necessary to implement the green transition on the scale proposed by the European Green Deal. These changes will have far-reaching and complicated effects, some of which may be unsavory. The stability of Europe and other vulnerable nations may depend on the details of these policies and how they are put into action.

In France, the gilets jaunes movement started as a protest against new carbon taxes, so it's clear that attempts to get people to choose renewable energy sources are polarizing (Figure 4.3).

Figure 4.3. *Protester in yellow vest on Paris street.*

Source: https://en.wikipedia.org/wiki/File:2018-12-29_14-58-30_manif-GJ-Belfort.jpg.

Critics have pointed out that the European Union's biofuel targets are driving up food prices and encouraging land conversion for crops like sugar cane and palm oil. Stranded assets and increased unemployment are two potential transition costs. The effects will be more noticeable in areas that used to be heavily dependent on carbon-intensive manufacturing.

Additionally, there is rising anxiety that certain delicate security-related outcomes may result from the global imperative to decrease carbon emissions. Nuclear power generation is being considered by some fragile nations as a means of producing low-carbon energy in the future, which has sparked concerns about the spread of nuclear technology. Another example is the potential emergence of so-called "green conflict minerals"—minerals that are essential for renewable energy and smartphone technology—due to the soaring demand for these commodities, which may inadvertently fuel violent conflict (Lazard and Youngs, 2021).

The most susceptible states to the effects of climate change are those that are fragile and have been hit by conflicts. Therefore, adaptation becomes even more crucial in these states. But these nations usually don't get much money for climate change. By facilitating future planning and offering a forum for discourse and trust-building among stakeholders, well-executed adaptation programs can bolster peacebuilding (Zwolski and Kaunert, 2011). However, because adaptation is a type of intentional social engineering, it has the potential to favor some members of society over others, ignore preexisting social structures, and even cause conflict if its dynamics are not foreseen.

For example: the European Union funded a climate adaptation initiative in Kenya that targeted wooded regions vital to the country's water supply. Human rights groups issued warnings and news of the Kenya Forest Service's violent removal of indigenous peoples from their homes in one of the project regions led to the project's 2018 suspension. The evictions claimed one life (Youngs, 2021).

4.3. CLIMATE SECURITY IN ASIA

Asia comprises a total of 51 countries. The region can be categorized into six subregions, taking into account their geographical location and proximity to the coast. In alphabetical order, we have:

- Central Asia with 5 countries;
- East Asia with 7 countries;
- North Asia with 2 countries;
- South Asia with 8 countries;
- Southeast Asia with 12 countries; and
- West Asia with 17 countries.

In 2023, the population of Asia was estimated to be around 4,753 million, accounting for approximately 60% of the global population. The population density was approximately 153 individuals per square kilometer. According to the CIA in 2023, Japan boasts the highest life expectancy at birth, reaching an impressive 85.03 years. On the other hand, Afghanistan has the lowest life expectancy at a mere 65.97 years. According to the World Bank, the gross domestic product (GDP) per capita varied significantly across countries. For instance, Afghanistan had a GDP per capita of $678 in 2023, while Singapore had a much higher GDP per capita of US$87,884.16 in 2023 (IGES, 2024).

4.3.1. Asia is Highly Climate-Vulnerable.

Asia has emerged as the largest regional-scale emitter of greenhouse gases globally, with China taking the lead in total yearly GHG emissions and India rapidly catching up. Nevertheless, the region is expected to encounter a disproportionate impact in the coming years, which paints a rather grim picture of its future prospects. According to some forecasts, the future success of Asia is at risk. According to Swiss Re's projections, the regional economy could potentially face significant challenges in the future due to climate-related costs, which are estimated to reach 26.5% by 2050.

Extreme weather events, such as heatwaves, floods, unpredictable monsoons, and rising sea levels, can significantly affect the region's economy and outdoor activities. There are numerous consequences that we must consider, such as:

- Significant human and economic losses;
- Destruction of ecosystems in the ocean and forests;
- Large-scale migrations; and
- The urgent threats of water shortage and food insecurity.

The impact of climate change and biodiversity loss in Asia extends beyond environmental concerns. It heightens the risk of agricultural damage due to invasive species and the emergence of new zoonotic diseases, such as COVID-19. Additionally, it paves the way for the wider dissemination of tropical diseases. The economic and financial channels, including growth potential, export receipts, public finances, and external debt, will face significant challenges in the coming decades due to the risks associated with increased political violence (Krampe and Mobjörk, 2018).

Asia's risk level is significantly higher due to its unique geographical characteristics. With a population surpassing any other nation, a significant portion of its inhabitants reside in densely populated urban regions that are prone to severe weather occurrences such as floods and heat waves. Additionally, many people live along extensive coastlines, making them more susceptible to cyclones and rising sea levels.

It is estimated that approximately one billion urban residents in Asia reside in areas with a high risk of flooding. The region exhibits a significant economic dependence on coal and oil, resulting in higher costs for transitioning to decarbonized economies. Despite this, it also demonstrates robust economic dynamism typically associated with rapid GDP growth and consequently, high energy consumption (IGES, 2024).

Climate risk indices emphasize the vulnerability of Asian countries by analyzing the occurrence and impact of previous natural disasters, as well as taking into account the concerning climate projections and specific country factors. Climate risks in the Asian region vary significantly, with different sub-regions experiencing varying levels of exposure and resilience. The vulnerability of different countries to weather-related events and their ability to adapt to extreme natural disasters is influenced by a range of factors. These factors include:

- Geography;
- Level of economic development;
- Demographic dynamics;
- The role of agriculture in the economy;
- Urbanization levels;
- Governance and institutions; and
- The strength of the domestic financial system.

Thus, based on the aforementioned factors, various groups of countries can be identified, each characterized by specific types of natural disasters and varying levels of intensity.

4.3.2. South Asia and the Bay of Bengal

South Asia and the Bay of Bengal region are home to countries that face significant climate security risks. Heatwaves pose a significant threat to South Asia, primarily because of the region's soaring temperatures and the prevalence of outdoor labor.

The Himalayan mountain chain is a matter of great regional concern. The accelerated melting of glaciers in this area will lead to devastating consequences for economic activity and the predominantly poor populations across multiple countries.

This includes floods and water scarcity, affecting 100s of millions of people. As per the World Bank, the issue of water shortage in South Asia is a significant concern that has the potential to impact over 1.5 billion individuals by 2050. This is a daunting prospect that, until recently, has not been adequately addressed by government policies (Krampe and Mobjörk, 2018).

Figure 4.4. *The worst floods in Pakistan's history have submerged more than a third of the country.*

Source: By Monica K. Smith, U.S. Army – http://www.defenseimagery.mil; VIRIN: 100811-A-8481S-884, Public Domain, https://commons.wikimedia.org/w/index.php?curid=11196882.

The Bay of Bengal, which encompasses eastern India, Bangladesh, and western Myanmar, is a region that faces significant climate risks due to its large coastal population. .

Figure 4.5. *Severe weather strikes Bangladesh.*

Source: By Rezowan at English Wikipedia – Own work (Original text: I (Rezowan (talk)) created this work entirely by myself.), GFDL, https://commons.wikimedia.org/w/index.php?curid=10555745.

These risks include sea level rises, with Bangladesh being the most vulnerable large country in the world in this regard, as well as intensifying cyclones, heavy rainfall, and prolonged droughts. Furthermore, the depletion of fish stocks and the destruction of mangroves pose a significant risk to food security in a region heavily reliant on agriculture. This vulnerability is compounded by the anticipated impact of natural disasters, such as unpredictable monsoons and limited water resources (Figure 4.5)

In addition, the sub-region will attract attention in terms of climate change due to geopolitical and conflict risks, as well as the negative impact of climate risks on social stability and ethnic tensions. These risks include its proximity to the strategic Strait of Malacca, where India, China, and the USA are vying for influence and maritime expansion. In the long term, it is anticipated that there will be mass migrations and increased risks of conflict (Prizzia and Levy, 2018).

4.3.3. India is Prone to Many Severe Natural Disasters

By 2025, India is expected to surpass all other countries and become the most populous nation in the world. Nevertheless, there are growing concerns regarding the potential consequences of climate change on the nation's future. It is a cause for concern, considering India's significant dependence on agriculture and related sectors. These sectors employ a substantial 44% of the population and makeup 20% of the country's GDP. Agricultural production is fraught with considerable risks, especially for the predominantly impoverished population (Prizzia and Levy, 2018).

India has witnessed a multitude of environmental events in the past few years, ranging from prolonged heatwaves and droughts to intense floods, powerful cyclones, and the rapid melting of Himalayan glaciers. In addition, it is important to note that the increasing sea level presents a substantial danger to the large population living along India's vast coastline. In the upcoming decade, Indian public authorities will be confronted with the task of finding a fine equilibrium between propelling the nation's progress and effectively handling the rise in greenhouse gas emissions.

India's role in climate change is increasingly significant, posing a potential threat to its development, despite its relatively lower per capita emissions compared to other major greenhouse gas emitters. Considering India's heavy reliance on coal, its high energy consumption, and its ambitious development goals, the country has made a firm pledge to attain "net-zero emissions" by 2070 during COP26, accompanied by a more substantial reduction in emission intensity by 2030 at COP27 (Krampe and Mobjörk, 2018).

Although the current targets may not have a substantial impact on reducing GHG emissions, it is likely that the country will be compelled to make more ambitious decarbonization commitments in the future as the consequences of climate change continue to escalate. There is a significant amount of importance involved, as emphasized by the thinktank ODI (Overseas Development Institute). It has been cautioned that in the absence of any action to address climate change, India may experience a significant annual decline in GDP ranging from 3%–10% by the year 2100.

4.3.4. Southeast Asian Agriculture and Food Security

Severe heatwaves, shortened monsoon seasons, and heavy rainfall will be challenges for the nations of South-East Asia, endangering economic activity, especially agriculture. These economies are susceptible to these climate risks because of their large outdoor, labor-intensive structures and small to moderate financial capacities.

Risks to food security are emphasized in the context of rice production. A major food staple in the area, rice is gaining attention because yields could be cut in half by 2100 if new, high-yielding varieties are not affected by climate change. Many low-lying nations in the subregion are also very concerned about the rising sea levels, particularly Indonesia, the largest archipelago in the world, which is moving its capital, Jakarta, to Nusantara, a future capital city that has not yet been built.

Figure 4.6. *Typhoon Haiyan hits Legazpi, Philippines with strong winds and waves.*

Source: By Lawrence Ruiz – Own work, CC BY-SA 4.0, https://commons.wikimedia.org/w/index.php?curid=43911464.

The countries surrounding the Mekong, the largest river in South-East Asia, are also experiencing an increase in climate risks. Recent years have seen significantly lower river flows and rainfall, which has further contributed to the decline in fish stocks. This will negatively impact agriculture and the economies of Cambodia, Laos, Myanmar, Thailand, and Vietnam, given the significance of the Mekong River.In the meantime, the Philippines is more vulnerable to stronger cyclones and warmer weather, which endangers fish and coral populations. Despite the fact that deforestation is slowing down in Indonesia and renewable power generation is becoming more popular, the ASEAN countries have not done much to prepare for these enormous risks up until now. Simultaneously, there appears to be no slowing down of the trend of rapid economic development driven by rising coal consumption (Figure 4.6).

4.3.5. Significant Risk and Substantial Economic Losses in the East Asia Region

Despite the significant exposure of East Asia to climate change, the overall impact seems to be somewhat mitigated due to the advanced state of its economies. In this sub-region of high-income economies, the robustness of macroeconomic fundamentals and the extensive financial systems will contribute to mitigating the impact of the climate bill. However, considering the current trajectory of GHG emissions, the economic costs will continue to rise at a rapid pace. One of the major concerns arises from the occurrence of droughts and water scarcity, as exemplified by the significant disruption faced by Taiwan's key chip industry in 2021. Extreme typhoons and floods pose growing threats, resulting in significant annual economic losses (Prizzia and Levy, 2018).

4.3.6. Climate Change Threatens China's Stability and Ambitions

China is a country that attracts a lot of attention due to its significant economic influence, the considerable economic impact of natural disasters, and its position as the top global emitter of greenhouse gases, although not on a per capita basis. Despite its significant financial resources and extensive global network, the country is vulnerable to the impacts of climate change. This vulnerability poses significant risks, including:

- High costs;
- Threats to food and water security; and
- Potential setbacks to the country's ambitious goals.

China has recently faced a series of challenging events, including severe floods in 2021, prolonged heatwaves and droughts in 2022, and historically low levels in the Yangtze River and other water supply sources. These natural disasters caused significant disruptions to energy supply from hydropower and industrial activity in central, eastern, and southern provinces (Krampe and Mobjörk, 2018).

When we think about the future, it's important to consider the potential for heatwaves, droughts, floods, and most importantly, water shortages in different parts of the country. We also need to be aware of the increasing risks of cyclones and the degradation of ecosystems throughout the entire nation. As a result, the pollution of soil and water reserves will have a significant impact on agriculture, infrastructure, supply chains, trade, and overall economic activity.

Xi Jinping's ambitious goal of achieving self-sufficiency for China may face challenges, as the country will still rely heavily on imports of grains and the development of climate-resilient crops. In this particular situation, China has made a commitment to reach the highest point of GHG emissions in 2030, achieve net-zero emissions by 2060, and internally, the country is striving for climate resilience by 2035 through a series of precautionary measures (IGES, 2024).

Even with the progress made in reducing the economic and human costs of natural disasters, the growing impact of climate change will only make this challenge more difficult. Given the current situation, it is worrisome to consider China's significant role in reversing the trajectory of greenhouse gas emissions. The potential impact of natural disasters on domestic power generation could potentially increase the dependence on fossil fuels in the coming years, further worsening the dynamics of climate change. There is another concern regarding the unchallenged political leadership of the Chinese Communist Party, considering the potential long-term impact of climate change on social stability (Krampe and Mobjörk, 2018).

4.3.7. Heightened Potential for Conflict and Social Unrest

In addition to the significant influence of climate change on the local economy, it is crucial to consider the political and social aspects when evaluating climate risks. These factors will contribute to increased social instability and conflict risks due to issues such as food and water insecurity and land losses in countries like Afghanistan, India, and Pakistan. The escalation of violent tensions and conflict in Asia will be driven by competing states vying for crucial access to water resources, resulting in two primary conflict risks. China's control over the upstream flows of major rivers in the region, such as the Mekong and Ganges, creates a potential conflict with its neighboring countries, including India.

In the event of reduced water volumes, China is likely to prioritize domestic use, especially considering the anticipated recurring droughts in the country. Furthermore, there is a longstanding dispute over the region of Kashmir between Pakistan and India, both of which possess nuclear capabilities. These two countries, with their large populations and significant climate risks, are currently vying for limited water resources. Increasing climate change will lead to significant migrations within and between states, which in turn will contribute to heightened tensions and risks of conflict. As per the IPCC, the occurrence of natural calamities in South Asia, South-East Asia, and East Asia resulted in the relocation of over 20 million individuals in 2019, and these numbers are anticipated to significantly increase in the future (Prizzia and Levy, 2018).

4.3.8. Mitigating the Accumulation of Additional Risks

Many countries in Asia are already taking action to address the challenges of climate change. Almost every country has made or updated commitments under the 2015 Paris Agreement, the significant global agreement on emissions reduction. China has recently announced its ambitious target of achieving carbon neutrality by 2060. Japan and Korea have both committed to achieving the same goal by 2050. However, there is still much work to be done to expand and expedite the shift towards a low-carbon economy. To achieve this ambitious goal, it is necessary to make significant changes in production and consumption patterns, as well as transform energy, transportation, and land use.

Implementing a carbon tax, which involves the government levying taxes on carbon emissions, has been recognized as a potentially powerful strategy for curbing CO_2 emissions (IGES, 2024). For example, Vietnam, a nation that has heavily depended on fossil fuels to fuel its rapid industrialization is also faced with numerous hazards due to its geographical location. Implementing a carbon tax of $25 per ton over the next decade would be a beneficial step towards achieving the country's Paris mitigation targets. Increasing the cost of carbon would motivate businesses and individuals to optimize their energy usage and promote a transition from coal-based energy to renewable sources. Carbon revenues amounting to approximately 1% of GDP could be allocated towards financing the nation's adaptation and mitigation strategies, as well as addressing various social development requirements.

Fiscal policy can play a crucial role in addressing the region's air pollution issue. Coal is responsible for a significant portion of emissions in China, India, and Mongolia, ranging from 68%–80%. These countries might want to consider implementing a specific tax on coal, with an equivalent carbon tax rate. India's coal tax, which was implemented in 2010 and later increased in 2020, has the potential for further enhancement. By implementing a coal tax equivalent to $25 a ton, a significant number of lives could be saved by 2030 in China alone (Krampe and Mobjörk, 2018).

An essential aspect of facilitating the shift towards a low-carbon economy will involve effectively addressing potential consequences, including increased energy expenses for individuals and businesses, job displacement, and uneven effects across different regions. However, the impact of policies will differ from one country to another. As an illustration, the implementation of a carbon tax would have varying impacts on different income groups in different countries.

In China and Mongolia, it would have a moderately regressive effect, meaning that it would be borne more by the poor. On the other hand, in India, it would have a moderately progressive effect, meaning that it would be borne more by the rich. It is crucial for countries with a regressive carbon tax to provide support for individuals, like coal miners, who rely on jobs in the energy sector. These workers often face financial challenges and may find it challenging to switch to emerging industries, such as renewable energy.

According to IGES (2024), to address the negative consequences of the transition, governments will need to explore methods of assisting households and businesses. In India, for instance, utilizing the revenue from a carbon tax to fund a universal lump-sum transfer (possibly utilizing Aadhaar unique identification numbers) would result in 80% of households experiencing improved financial situations and a reduction in inequality. In China, implementing both a universal lump-sum transfer per person and a subsidy to rural households could be effective in addressing inequality. Displaced workers employed in affected sectors could receive support through extended unemployment benefits, training, and reemployment services. Additionally, increasing public spending, such as investing in clean public infrastructure, has the potential to generate employment opportunities in low-carbon industries. It would be beneficial for governments to implement market-based incentives that encourage the availability of green finance, thus alleviating financial burdens for companies (Prizzia and Levy, 2018).

4.4. CLIMATE SECURITY IN AFRICA

The connections between climate change and insecurity are intricate and cumulative. Temperature fluctuations, variations in rainfall patterns, changes in soil moisture levels, and alterations in water acidity levels can all act as "threat multipliers" for political and social violence. As access to food and water becomes more limited, conflicts can escalate among farmers, herders, people living in rural areas, and city dwellers.

The Horn of Africa is confronted with a multifaceted crisis encompassing climate change, political instability, and profound under-development. The problem is being worsened by recurring droughts and sudden floods, as well as by rising temperatures and growing competition for limited arable land. Climate change is exacerbating migration and displacement, escalating conflicts over water and grazing, and increasing the likelihood of organized violence in Ethiopia, Eritrea, Somalia, Sudan, South Sudan, and neighboring nations (Garcia, 2008).

Over the past decades, there has been a significant increase in the understanding of the connections between climate change, peace, and security. Africa, in particular, has been extensively researched on this subject, making it one of the most well-studied regions in the world. Simultaneously, there has been significant attention given to addressing climate-related security risks in Africa, allowing for the identification of emerging responses and best practices (Daoudy, Sowers, and Weinthal, 2022).

4.4.1. The Complexity of Conflicts in Natural Resource Management

A pressing concern in Africa is the escalating competition for natural resources, which poses a significant climate security challenge. This is especially worrisome because a significant number of Africans depend on these resources for their food and livelihood security. Climate impacts are intensifying the current strains on natural resources like land, water, and forests, and are changing the way we can access and use these resources. The impacts of climate change on natural resources vary across the continent. When there is widespread availability of small weapons, coupled with a long history of conflict and social divisions, marginalized groups, and inadequate institutions for managing resources and resolving disputes, this competition has the potential to escalate into violent conflict (UNEP – UN Environment Programme, 2024).

First, while it is widely known that conflicts between farmers and herders have been extensively studied, recent research suggests that natural resource conflicts are much more complex. These conflicts involve multiple groups, resources, and sectors, such as fisheries, mining, and conservation. Conflicts and tensions surrounding water, fisheries, forests, and land are intensifying throughout Africa (Garcia, 2008).

Second, in regions that already experience political and social instability, the chances of violence in conflicts over natural resources tend to be higher. Exclusion and marginalization often have a profound effect, on existing divisions within society (Brown, Hammill, and McLeman, 2007). Governance plays a vital role as a mediator. If a system or approach is functional, thorough, and within the bounds of the law, it has the potential to successfully prevent conflicts and peacefully manage increasing pressures. Poor governance increases the potential for risks to escalate.

Third, most conflicts concerning natural resources tend to arise at a sub-national and local level. As a result, the way they interact with each other is greatly influenced by the specific local conditions and differs from place to place (Daoudy, Sowers, and Weinthal, 2022). In certain regions of Eastern Africa, the main form of conflict among pastoralists revolves around the act of stealing cattle, commonly referred to as cattle raiding. Fourth, competition among states for natural resources, particularly transboundary water resources like the Nile, Juba-Shabelle, and Zambezi River Basin, is increasing. The current political tensions among states may escalate as a result of the collective pressures stemming from economic development, population growth, and climate impacts. Furthermore, when upstream states make decisions that unilaterally affect downstream users, the potential for escalation rises.

4.4.2. Pressures on Livelihoods, Food, Water, and Energy

Security of food, water, and energy as well as livelihoods are under threat throughout the continent. The effects of climate change are making other problems with food, water, and energy security worse. As of right now, 20% of Africans experience hunger, 1.34 billion lack access to clean, affordable energy, and many lack access to both (Daoudy, Sowers, and Weinthal, 2022). A multitude of security-related issues can be fueled by livelihood and food, water, and energy insecurity, especially in precarious circumstances.

Interruptions in the supply of food, water, and energy, as well as sudden price increases, can exacerbate pre-existing complaints and serve as catalysts for political unrest and demonstrations (Garcia, 2008). Protests over food prices occurred in all African countries during the last two global food price crises

Furthermore, there is a growing number of weather-related disasters in the continent. They worsen instability risks and further strain livelihoods, particularly when disaster responses are seen as unjust or inadequate (Brown, Hammill, and McLeman, 2007). For instance, after Cyclone Idai in 2019 left the affected populations without enough food, energy, or water, protests took place in Mozambique and Zimbabwe (Figure 4.7).

Figure 4.7. *Kenyans protest high inflation, living costs, and food prices.*

Source: (https://commons.m.wikimedia.org/wiki/File:Kenya_2024_protests_%289%29.jpg).

Natural disasters are becoming more frequent and expensive, which makes it more difficult for governments to fund development and offer essential services. This can erode political legitimacy and bolster anti-state sentiments.

Insecure livelihoods can lead to detrimental coping strategies and maladaptation, such as:

- Mining;
- Illegal fishing;
- Deforestation; and
- Poaching.

Armed groups utilize some of these activities as a means of self-financing. For instance, Al Shabaab controls taxes and earnings from the production of charcoal in Somalia, and the Rapid Support Forces (RSF) in Sudan rely heavily on the proceeds of artisanal gold mining.

4.4.3. Climate Change and the Movement of People

For centuries, human mobility has played a crucial role in the way of life for numerous communities across Africa. This aspect continues to play a crucial role in sustaining livelihoods and effectively dealing with changes. Climate impacts are becoming more influential in these dynamics and are speeding up several mobility trends. When movements are forced, sudden, at a larger scale, irregular, and/or unplanned, the risks for both migrants and host communities increase. The pressures arising from demographic changes related to mobility can often lead to tensions and conflicts. When migration is properly managed and coordinated, it continues to be a crucial factor in driving economic development and resilience.

The most significant migration trend affected by climate risks is the movement within countries, specifically from rural areas to cities. The impacts of climate change are putting rural livelihoods at risk, leading to an increase in rural-urban migration. It is projected that by 2050, there will be an estimated 87.5 million internal migrants (Garcia, 2008).

-When cities struggle to handle the sudden influx of new migrants and fail to provide adequate services and opportunities, it can lead to tensions between different population groups and an increase in grievances (Brown, Hammill, and McLeman, 2007). In light of this context, there is a noticeable rise in cities that are rapidly expanding and becoming climate security hotspots throughout the continent.

The frequency and severity of weather-induced disasters is on the rise, leading to an increase in the risks of displacement. Certain groups, including IDPs and refugees, women, children, people with disabilities, and minority groups, are more susceptible to facing significant risks. Challenges frequently arise in receiving areas, encompassing water, energy, and food insecurity, along with conflicts over access to natural resources, public services, and jobs (Daoudy, Sowers, and Weinthal, 2022).

According to a report by the Africa Center for Strategic Studies in 2023, cross-border migrants tend to remain within the same region. This migration is commonly attributed to economic factors, but climate risks are also playing a significant role by impacting people's livelihoods. Migration patterns often involve individuals relocating to urban areas in nearby countries, which can exacerbate the existing pressures and challenges faced by cities.

The escalation of tensions and risks is often observed when xenophobia rises and politicians exploit migrants as scapegoats for economic, social, and political challenges. This phenomenon is evident in regions like North and Southern Africa.

Migration from the African continent to other continents is relatively smaller compared to mobility within the continent. However, the increasing securitization of borders in countries within and outside of Africa has limited the legal channels available for international migrants. As a result, they are forced to take more dangerous routes, putting them at a higher risk of encountering smugglers and traffickers. International migrants are also exposed to considerable risks when it comes to experiencing abuse. It is worth noting that there is a higher risk for women and children (Garcia, 2008).

4.4.4. Non-State Armed Groups Take Advantage of Climate Security and Governance Issues

The security risks associated with climate change are providing a favorable environment for non-state armed groups, including terrorists, criminal organizations, and militias, throughout Africa. The effects of climate change are influencing different groups' strategies. These organizations fill the gap and provide necessary services and governance in areas where state institutions are weak and corrupt. The effects of climate change on livelihoods are also encouraging these groups to recruit new members and increase the likelihood that they will use natural resources as weapons.

The effects of climate change are putting more strain on state institutions. Nonstate armed groups have the ability to erode state institutions even more and establish their own authority and power in circumstances where state institutions are unable or unwilling to rule and provide essential services. All throughout the continent, the dynamics described are apparent. Maystadt and Ecker (2014) draw attention to Al Shabaab's provision of financial assistance and disaster relief in Somalia in the aftermath of droughts.

The extremist organization Ahlu Sunnah wa-I-Jama'ah (ASWJ) has attempted in the past to build credibility in the marginalized areas of northern Mozambique by conducting proselytizing and offering their own religious instruction. The different elements that lead to armed group recruitment—such as limited social and economic opportunities, insecurity in one's livelihood, and the financial incentives provided by armed groups—are made worse by the effects of the climate.

The African continent as a whole is experiencing the described trend. Many young people in the Lake Chad region view armed groups as a means of escaping their difficulties, particularly those who have been displaced or have lost their sources of income (UNEP – UN Environment Programme, 2024).

Young pastoralists in Eastern Africa also face similar circumstances, particularly those who suffer from severe political marginalization and unjust treatment. The Intergovernmental Authority on Development's Conflict Early Warning and Response Mechanism (2023) draws attention to this.

Similar to this, the perceived benefits of joining armed groups in Central Africa—such as improved socioeconomic prospects and protection of their livelihoods, which include farms and cattle—have drawn young people from rural areas to join them.

Environmental problems are usually made worse by armed groups. These organizations typically rely heavily on illegal resource extraction and environmental crimes like poaching, mining, and logging as major sources of funding. Concern is being raised by the events occurring throughout the continent, particularly in the Congo Basin. One of the last significant net carbon sinks in the world is at risk due to these activities **(Garcia, 2008)**.

In terms of net forest loss from 2010–2020, for example, the DRC ranks second only to Brazil and ahead of Indonesia with a deforestation rate of 0.83% annually (UNDP Climate Promise, 2023).

Armed groups are increasingly considering the effects of climate change when making strategic decisions that influence their operational strategies. The use of water infrastructure as a weapon in the exploitation of natural resources is one instance. For example, communities in Libya have been forced to drill private wells as a result of militant groups' persistent sabotage of water pipes, which has made the problem of water scarcity worse (Brown, Hammill, and McLeman, 2007).

Climate-related risks can also make it more difficult for armies and peace-keeping missions to carry out their tasks, particularly if they are unable to predict extreme weather. Floods in South Sudan have made it difficult to provide aid and protect civilians, so the peace mission has had to improve its capabilities and modify its operations in order to be ready for severe weather (Figure 4.8).

Figure 4.8. *Children in Sudan navigate floodwaters.*

Source: By Ashorooq Tv قناة الشروق الفضائية *– https://www.youtube.com/watch?v=HZ0-OmgqcFY, CC BY-SA 4.0, https://commons.wikimedia.org/w/index.php?curid=128995542.*

4.4.5. Africa Faces Geopolitical Security Risks Associated with Climate Change.

Security risks associated with climate change in Africa are heavily contingent upon and shaped by multiple external factors. Africa's capacity to adjust to climate shocks is shaped by geopolitics and policies in other parts of the world, particularly in relation to food systems, energy, trade, migration, and extractive industries. Important conclusions

African food systems are vulnerable due to their high reliance on imports and international markets, particularly given the region's comparatively low adoption of climate-smart agriculture and low domestic production of essential foods. Supply shocks may result from global crises in food prices (Garcia, 2008).

Because international fishing can have a detrimental effect on local livelihoods and thereby fuel security-related problems like piracy and trafficking, it reduces the climate resilience of coastal communities. For instance, piracy has increased in Somalia as a result of a combination of droughts and an increase in foreign industrial fishing. While multinational corporations can be crucial in creating socioeconomic opportunities, they can also exacerbate climate security threats, particularly when they participate in rent-seeking and corrupt activities that worsen the environment (Garcia, 2008). They may take on the role of conflict actors in situations involving extremely delicate subjects like mining and the management of water and land. For instance, private logging companies have historically contributed to violence and conflicts in Liberia due to their activities and earnings, which have been connected to political elites and multinational corporations (Brown, Hammill, and McLeman, 2007).

30% of the world's mineral reserves are thought to be found in Africa; these reserves are essential for the continent's green transition and, as demand grows, present both opportunities and risks. In the majority of these countries, the mining industry still faces social and environmental challenges and primarily provides income for a small number of elites, without contributing significantly to local development or sustainable growth. For instance, the mining industry in many Central African economies is poorly regulated and continues to be one of the main causes of conflict as well as human rights abuses and violations, despite its significance for global trade (Daoudy, Sowers, and Weinthal, 2022).

Attempts to restrict migration, rather than providing legal pathways, continue to be the primary driving force behind migration policies in many non-African nations. These policies exacerbate the vulnerabilities that migrants face by raising the risks for those who need or want to migrate and by indirectly supporting criminal networks that benefit from irregular migration (UNDP Climate Promise, 2023).

4.4.6. Climate Change Poses Multiple Security Threats.

Climate-related security risks are influenced by various individual risk factors, such as:

- Gender;
- Age;
- Disability; and
- Socio-economic status.

The experience of marginalization, inequality, and exclusion varies depending on how these factors intersect.

Generally, individuals who are more vulnerable tend to be more affected by climate-related security risks.

The compounding impacts of climate change tend to disproportionately affect societal groups that have historically been marginalized. These groups, including women, young people, children, IDPs, and refugees, persons with disabilities, and other minorities, often face barriers to accessing the resources, opportunities, or services needed to cope and adapt (Daoudy, Sowers, and Weinthal, 2022). Women often face greater vulnerability to natural hazard-induced disasters, especially when it comes to personal safety and gender-based violence (GBV). Due to prevailing patriarchal norms, women across the continent are frequently left out of land rights and decision-making regarding resource management. In addition, the labor-intensive household roles that women typically have can create additional challenges for them in terms of expanding their livelihood options or building resilience to climate shocks (Brown, Hammill, and McLeman, 2007).

Climate change has wide-ranging impacts on various aspects of society, including:

- Food and livelihood security;
- Mental and physical health;
- Human mobility;
- Decision-making and governance; and
- Criminality and armed group recruitment.

4.4.7. Tackling the Security Threats Posed by Climate Change in Africa.

To effectively tackle the challenges in Africa, it is necessary to develop solutions that are based on the continent's own context and circumstances. The security risks associated with climate change that African states and societies face are largely similar, but their specific manifestations and dynamics are invariably very localized. The way political, traditional, and corporate leaders behave, the underlying political economy and power structures, governance structures and institutions, histories of conflict, marginalization, and identity, and all of these factors play a critical role in converting climate risks into security and conflict risks. These variables are very contextual and vary greatly between nations, regions, and local communities. This means that the following areas need special attention to address security risks related to climate change:

1. **Enhance Data Collection, Research, and Analysis:** African analysis, research, and data collection capacities must be strengthened to provide a more contextualized analysis of how climate-related security risks manifest to inform appropriate solutions.
2. **Make Use of Customary and Indigenous Wisdom and Establishments:** Utilizing local, traditional, and indigenous institutions and knowledge is crucial to the success of interventions aimed at addressing climate security, and it ought to play a major role in an integrated strategy.
3. **Leverage the Youth's Potential:** Equipped with appropriate resources, they can catalyze swift advancement and creativity. Jobs and education will be essential for this to become a reality.
4. **Do Not Abandon African Island States:** They require tailored solutions, funding, and capacity building. These include improving

maritime awareness and enforcement capabilities, allocating resources for early warning and disaster relief, formulating a comprehensive strategy for food, energy, and water security, and holding frequent discussions and knowledge-sharing forums.

4.4.7.1. Funding Initiatives Related to Climate Security

Access to financing will be essential for addressing climate-related security threats across the board and for creating more resilient and sustainable economies, societies, and states throughout Africa. There are several particular areas in which immediate action is required:

1. **Close the Climate Financing Gap:** More money must be spent immediately on risk prevention and resilience building to avert the worst-case scenarios. This covers investments in absorption capacities as well as easier and better access to finance for adaptation. In addition, to make up for the harm that climate change has already caused, an international mechanism for loss and damage must be operationalized.
2. **Reach the Most Fragile, Conflict-Affected, and Vulnerable Contexts:** Strengthening capacities to manage these funds is necessary, particularly at the national level, and financing for these contexts needs to be increased and made easier to access.
3. **Strengthen African Financing Facilities:** The African Peace Fund should be rapidly operationalized and scaled up to address climate security risks on the continent, and access to the African Risk Facility should be made better.
4. **Offer Integrated Financing to Advance Climate Security Actions:** The first experiences with tackling climate-related security risks that have already been gained, along with the emerging best practices that are detailed in this report, should be developed and further upgraded with significant funding schemes for climate security.

4.4.7.2. Institutions, Collaboration, and Resilience-Building Prevention

There will be a significant increase in climate-related security risks in the coming years and decades. To mitigate the most severe consequences, it is crucial to shift from a reactive, responsive strategy to a proactive, forward-thinking approach that prioritizes prevention, resilience building, and sustainable development on a larger scale. To make this possible, it would be helpful to consider the following areas as significant entry points:

1. **Integrating Climate Security Into Strategies and Policies:** It is crucial to ensure that climate, peace, and security considerations are fully integrated into all relevant strategic and policy frameworks. This includes:
    a. National adaptation plans;
    b. Nationally determined contributions;
    c. Climate change action plans;
    d. Peace plans;
    e. Interventions and architectures;
    f. Defence strategies;
    g. Stabilisation frameworks; and
    h. Sectoral policies such as water, agriculture, and energy.
2. **Climate and Conflict Must be Linked in Early Warning and Action Systems:** These systems should explicitly consider the indirect and cascading risks associated with climate change and environmental stress. This can be achieved by incorporating a comprehensive set of climate security indicators and by building upon local institutions, indigenous knowledge, and civil society.
3. **Encourage Collaboration Across Different Sectors:** The objective is to enhance operational responses, specifically by incorporating climate considerations into peacebuilding and conflict prevention efforts. In order to facilitate such cooperation, institutions must have effective leadership and provide appropriate incentives.
4. **Organizations Should Focus on Developing and Strengthening Their Climate Security Capacities and Institutions:** This can be achieved by having dedicated climate security specialists and establishing effective cooperation and coordination structures, both vertically and horizontally. It is crucial for organizations and institutions at every level to establish transparent lines of responsibility and accountability. To promote these advancements, it is crucial to prioritize the exchange of experiences, training, and capacity building in this field. One potential solution could be the establishment of an AU-led training facility dedicated to climate security.
5. **African Cities Have the Potential to be Powerful Agents of Change:** It is crucial for them to tap into their human resources, address the growing issues of inequality and exclusion, and create a supportive atmosphere for sustainable development. It is crucial to prioritize enhancing cross-sectoral cooperation to promote localized

solutions that effectively tackle social, economic, environmental, and political challenges holistically. This can be achieved by implementing inclusive urban planning and development initiatives (Garcia, 2008).

4.4.7.3. Resilience Characterizes a Robust Civil Society.

The forefront of addressing security risks associated with climate change is civil society. Civil society actors play a crucial role in building stronger societies and states, especially in vulnerable and conflict-ridden areas. They serve as advocates and catalysts for the necessary transformations. To effectively mitigate security risks associated with climate change, it is imperative to focus on the following areas:

1. **Develop Awareness and Capacities:** To facilitate civil society action on security risks associated with climate change, it will be critical to increase awareness, offer best practices, and fund training and capacity building. Increasing the capacities of civil society to address climate security will be essential to close the action gap on these risks at the local level.
2. **Implementation Partner:** Using the assets of civil society—its networks and implementation structures, its experience implementing projects in difficult environments, its analytical skills and knowledge of the local context—partnerships for implementing climate-security risks should be upgraded to guarantee delivery on the ground.
3. **Safety for Individuals at the Forefront of Climate Security Threats:** Human rights advocates and environmental defenders require increased involvement, education, and safeguards. Investments in environmental defenders' capacity, agency, and protection should be given top priority.
4. **Involve the Private Sector:** Through collaborations with the private sector, civil society can facilitate the private sector's constructive involvement in mitigating security risks associated with climate change. This can be achieved, for instance, by guaranteeing that infrastructure projects are conflict- and climate-sensitive and that they yield tangible advantages for nearby communities.

4.5. CLIMATE SECURITY IN OCEANIA

It is widely accepted in the scientific community that global warming is leading to significant changes in the Earth's climate system, which in turn has various impacts on environmental security. First and foremost The Pacific Islands share similar geographic characteristics, which make them susceptible to the environmental effects of climate change. Continental states do not face the same exposure to tropical storms and rising sea levels as they do. Reef-generated fisheries and tourism, crucial for island nations, face significant threats from ocean acidification and rising temperatures. One of the challenges we face with fresh water supplies is their vulnerability. This vulnerability stems from the limited land area available to capture precipitation and the risk of groundwater exposure to saline intrusion caused by rising sea levels (Campbell, 2019).

According to the latest assessment report from the Intergovernmental Panel on Climate Change (IPCC), small islands face significant risks in the coming decades. Soon (2030–2040), there is a projected medium risk of losing livelihoods, coastal settlements, infrastructure, ecosystem services, and economic stability. Looking further ahead to the long term (2080–2100), the risk is expected to be very high. In a broader sense, it predicts a decline in biodiversity, fish populations, and the protective capabilities of coral reefs. It also anticipates increased flooding and habitat loss caused by rising sea levels, extreme weather events, shifts in precipitation patterns, and a decrease in ecological resilience. These risks are deemed high in the near future and even more severe in the long run (Campbell, 2019).

4.5.1. Considering the Security Implications

One of the significant security concerns for the Pacific Islands revolves around climate-related issues. These include:

- The availability of fresh water, which is affected by changes in rainfall patterns and salt water intrusion.
- Concerns about the local food supply, as damage to coral reefs, declining fisheries, and impacts on agriculture pose challenges
- Infrastructure damage is a pressing issue due to rising sea levels, flooding, and storm damage.

There are several potential second-order consequences that could arise from these events. These include:

- Economic loss;
- Declining revenues from tourism; and
- An increase in emigration as people try to escape the situation, particularly from atoll islands that are at risk of being inundated due to sea level rise. For certain Island nations made up entirely of low-lying atolls, such as Kiribati, Tuvalu, and the Marshall Islands, the rise in sea levels poses a significant existential danger.

The anticipated impacts of climate change are interconnected with various global trends, including population growth, water, and air pollution, and the rising demands for natural resources, like tuna, from developing nations. According to the U.S. Department of Defense, climate change will have significant security implications. It will not only impact resource competition but also place additional burdens on economies, societies, and governance institutions globally (Boege, 2018). These effects can greatly exacerbate stressors in other countries, such as:

- Poverty;
- Environmental degradation;
- Political instability; and
- Social tensions.

These conditions have the potential to facilitate terrorist activity and other forms of violence. Pacific Island nations, as highlighted by the Pacific Islands Forum (PIF) Secretariat, acknowledge the gravity of the security threat posed by climate change. They emphasize that climate change poses an immediate and significant challenge to sustainable development, poverty eradication, and even the survival of some Pacific Island Countries. Due to their geographical location and position in the middle of the ocean, they find themselves at the forefront. However, these countries have a limited capacity to adapt and respond, resulting in consequences that are unfairly large compared to their minimal contributions to global emissions (Boege, 2018).

4.5.2. Addressing the Security Impacts of Climate Change in the Pacific Islands

Managing climate change security impacts involves utilizing tools such as:

- Adaptation and response to climate change;
- Greenhouse gas mitigation; and
- Knowledge creation.

Mitigation involves implementing measures to decrease the levels of greenhouse gases in the atmosphere, which consequently affects the oceans as well.

Activities aimed at enhancing resilience to climate change impacts will be determined by the effectiveness of global mitigation strategies. Response encompasses efforts aimed at mitigating and adapting to higher levels of danger, disaster, or catastrophe, even when our best efforts to prevent them fall short. Understanding and generating knowledge is crucial for addressing the intricate and evolving challenges posed by climate change (Campbell, 2019). It allows us to:

- Take proactive measures;
- Adapt to changes; and
- Respond effectively.

Specifically, the security sector requires practical knowledge to inform policy and planning. There are several regional organizations that are involved in cooperative initiatives. These include:

- The Pacific Islands Forum (PIF);
- Melanesian Spearhead Group (MSG);
- Secretariat of the Pacific Community (SPC);
- Secretariat of the Pacific Regional Environment Program (SPREP).

4.5.2.1. Mitigation

The nations and territories of the Pacific Islands make up very little of the world's greenhouse gas emissions. Despite this, there is a general awareness of the region's reliance on fossil fuels, particularly in areas where diesel generators are the primary source of electricity and where maritime goods transportation is essential to commerce. The Majuro Declaration in September 2013 highlights the shared policy emphasis among island states on the adoption of renewable energy for sustainable development. Presenting a unified front to the COPs to the UN Framework Convention on Climate Change (UNFCCC) in their endeavor to forge a binding global accord on greenhouse gas emissions during the 2015 Paris meeting was a fundamental political goal of the Majuro Declaration. Since the Kyoto Protocol was drafted in 1997, the parties have failed to achieve this goal. As a result, the Majuro Declaration is the result of discussions between the major emitters and Pacific Island nations. The declaration indicates a regional intention to cut greenhouse gas emissions, but national commitments underpin it. A declaration's appendix compiles particular pledges made by the 15 PIF members (Campbell, 2019).

4.5.2.2. Adaptation

The UNFCCC established a fund to assist in the development and execution of national adaptation plans of action (NAPAs), offering the "least developed countries" (LDCs) substantial support in planning for climate adaptation. Kiribati, Samoa, the Solomon Islands, Tuvalu, and Vanuatu are Pacific Island countries that have finished their NAPAs and are receiving assistance from the LDC Fund (Boege, 2018).

The World Bank has committed to increasing its investments and has given $140 million for policy support and climate adaptation projects in six Pacific Island countries. Using SPC and SPREP, the Global Climate Change Alliance

(GCCA) of the European Union has funded adaptation initiatives in nine Pacific Island nations. However, the Pacific Islands receive a large portion of their funding for climate adaptation and mitigation from bilateral aid and technical assistance.

According to the Organization for Economic Cooperation and Development, global bilateral aid related to climate change amounted to $21.5 billion between 2010 and 2012. Out of this total, the majority, 58%, was allocated towards mitigation efforts, while 25% was directed towards adaptation initiatives. The remaining 18% was utilized for mixed purposes. 45% of aid went toward climate adaptation projects in Small Island Developing States (SIDS) across the globe. While the total amount of adaptation aid received by nations in Oceania was only 2%, on a per capita basis SIDS, which includes the Pacific Island nations, received the most of this aid.

The UNFCCC Adaptation Fund and organizations like the World Bank and Asian Development Bank have been providing multilateral aid to Pacific Island countries, and this trend is anticipated to continue. For instance, a financing agreement worth EUR 37 billion was inked in February 2014 by the European Union and PIF for climate change adaptation and sustainable energy in the Pacific Islands. SPREP is striving to secure and oversee funding from the UNFCCC's Green Climate Fund. However, there hasn't been much of an overall advancement toward climate resilience (Boege, 2018).

Many $100 million in climate-related aid have reportedly been sent to the Pacific Islands over the past 25 years, according to a recent UN Development Program (UNDP) report. Nevertheless, the report highlights that many of these initiatives have not effectively communicated the importance of long-term sustainable adaptation to Pacific Island communities, nor have they successfully implemented and integrated suitable solutions across the region.

Each of the four regional organizations that were previously mentioned is working on multilateral climate adaptation projects. For instance, PIF oversees cooperative climate initiatives with the US and the EU, and it spoke about climate issues at the 2013 Majuro Summit. Among the many adaptation, mitigation, and policy development initiatives that SPREP works on are the Pacific Adaptation to Climate Change Project and its collaboration with UNDP, which manages national initiatives across 14 Pacific Island nations and territories.

In 2013, MSG appointed an officer to oversee the environment and climate change, and it has completed a memorandum of understanding with Griffith University in Brisbane, Australia, for climate change projects. SPC oversees several climate adaptation projects supported by the US, Germany, and the EU and has written policy briefs on related subjects (Campbell, 2019).

4.5.2.3. Response

Considering the security sector's viewpoint, climate response is seen as a crucial aspect of humanitarian assistance and disaster relief (HA/DR). According to scientific consensus, it is expected that climate change will lead to a rise in natural disasters. Consequently, there will be an increasing demand for humanitarian assistance and disaster relief throughout this century. The planning for that challenge is still in its early stages, as shown by a special session on "Climate Change, HA/DR, and Security in the Asia Pacific," which took place on May 31, 2014, as part of the Shangri-La Dialogue (Boege, 2018).

4.5.2.4. Creation and Distribution of Knowledge

Given the ever-present threat of climate change, it is crucial to constantly acquire new knowledge to effectively plan for mitigation, adaptation, and response. Due to the efforts of the IPCC, there has been a significant increase in the understanding and accessibility of scientific knowledge regarding climate change, including its environmental effects and the societal consequences it entails. This information is now widely available to both governments and the general public.

Regional organizations are actively involved in the development of knowledge networks for climate adaptation and response, often working together with partner nations. During the period from 2010–2014, the University of the South Pacific took charge of the Pacific component of the Global Climate Change Alliance, which was financially supported by the European Union. This project aimed to enhance the skills of climate adaptation professionals and promote practical research in Pacific Island nations. In recent times, SPREP has collaborated with Griffith University, with the backing of the Australian government, to carry out a Pacific Climate Change Information Management project. There are various research and knowledge dissemination programs in the United States that focus on the Hawaiian archipelago and the U.S.-affiliated Pacific Islands. As an illustration, the Pacific Islands Regional Climate Assessment (PIRCA) has been funded by the Department of the Interior and the National Oceanographic and Atmospheric Administration (NOAA). It has conducted assessments on climate change indicators and adaptive capacities of these islands and has contributed valuable input to the 2014 U.S. National Climate Assessment. A Pacific Island Climate Science Center is managed by a consortium consisting of the University of Hawaii Manoa, the University of Hawaii Hilo, and the University of Guam. The center is funded by the Department of the Interior. Additionally, NOAA has developed a web-based Pacific Climate Information System (PaCIS) to offer valuable information for climate adaptation purposes (Boege, 2018).

CHAPTER 5

INTERNATIONAL COOPERATION ON CLIMATE SECURITY

Contents

INTRODUCTION

Governments around the world have made commitments to mitigate climate change for decades. However, the world is already experiencing the effects of climate change, and these are only going to worsen, regardless of how much diplomacy is being conducted.

The Paris Agreement and the Kyoto Protocol committed nations to lower their emissions of greenhouse gases; however, atmospheric carbon dioxide levels are continuing to rise, rapidly causing the planet to warm. Environmental disasters, including unprecedented sea level rise, record-breaking droughts and floods, and massive extinctions of species, could befall much of the globe if current warming trends persist unchecked.

Many of the 195 signatory nations to the Paris Agreement have increased their climate pledges since its 2015 negotiations, at the United Nations Climate Conferences, called the Conference of the Parties (COP), by, for example, promising to reduce emissions and provide aid to nations struggling to adapt to the consequences of severe weather.

Below are some of the most significant international climate change agreements such as:

1. **Montreal Protocol:** The Montreal Protocol was established in 1987. This was a significant environmental agreement that set a precedent for future diplomatic efforts in addressing climate change. It is worth noting that the treaty was ratified by every country in the world, signifying their commitment to cease the production of ozone-depleting substances, including chlorofluorocarbons (CFCs). The protocol has achieved an impressive reduction of almost 99% in these ozone-depleting substances. In 2016, parties reached an agreement through the Kigali Amendment to additionally decrease their production of hydrofluorocarbons (HFCs), which are potent greenhouse gases that contribute significantly to climate change (Velders et al., 2007).
2. **The UN Framework Convention on Climate Change (UNFCCC):** This was established in 1992. The landmark accord, ratified by 197 countries, including the United States, stands as a significant global treaty that specifically tackles the issue of climate change. An annual forum, called the Conference of the Parties, or COP, was established to facilitate international discussions to stabilize the concentration of greenhouse gases in the atmosphere. During these meetings, significant international agreements were reached, including the Kyoto Protocol and the Paris Agreement.
3. **The Kyoto Protocol:** This was established in 2005. The Kyoto

Protocol, which was adopted in 1997 and came into effect in 2005, stands as the pioneering climate treaty with legally binding obligations. Developed countries were mandated to decrease emissions by an average of 5% below 1990 levels, while a monitoring system was put in place to track their progress. However, it is worth noting that the treaty did not impose any obligations on developing countries, such as China and India, who are significant contributors to carbon emissions, to implement any measures. In 1998, the United States signed the agreement, but unfortunately, it never followed through with ratification and eventually decided to withdraw its signature.

4. **The Paris Agreement:** This was established in 2015. The Paris Agreement, which is considered the most significant global climate agreement thus far, mandates that all nations establish commitments to reduce emissions. Governments establish targets, referred to as nationally determined contributions (NDCs), to prevent the global average temperature from surpassing the 2°C (3.6°F) mark above preindustrial levels. Additionally, they strive to make endeavors to ensure that the temperature remains below 1.5°C (2.7°F). The ultimate goal is to achieve global net-zero emissions by the second half of the century. This means that the amount of greenhouse gases emitted will be equal to the amount removed from the atmosphere. (This concept is commonly referred to as achieving climate neutrality or carbon neutrality.)

In November 2020, the United States, being the world's second-largest emitter, made the decision to withdraw from the accord. This move was initiated by former President Donald Trump. President Joe Biden, in his first months in office, decided to reenter the United States into the agreement. It is worth noting that the agreement has not been formally approved by three countries: Iran, Libya, and Yemen.

5.1. MONTREAL PROTOCOL

The Montreal Protocol is a significant international agreement that governs the manufacturing and usage of approximately 100 artificial chemicals known as ozone depleting substances (ODS) which harm the ozone layer. Upon being released into the atmosphere, these chemicals inflict harm upon the stratospheric ozone layer, which serves as Earth's safeguard against excessive ultraviolet radiation from the sun, shielding both humans and the environment. Adopted on September 16, 1987, the Protocol is currently one of the few treaties that has been universally ratified. **The Montreal Protocol gradually reduces the use and manufacturing of various ODS through a systematic approach, with** separate schedules for developed nations and developing nations (referred to as "Article 5 countries"). According to this treaty, all parties have distinct obligations regarding the gradual elimination of various categories of ODS, regulation of ODS trade, regular submission of data, implementation of national licensing systems to monitor the import and export of ODS, and other related issues. Both developing and developed countries have equal but differentiated responsibilities. Crucially, both groups of countries are obligated to make binding, time-targeted, and measurable commitments (Mäder et al., 2010). The Protocol encompasses provisions pertaining to:

- Control Measures (Article 2);
- Calculation of control levels (Article 3);
- Control of trade with non-Parties (Article 4);
- Special situation of developing countries (Article 5);
- Reporting of data (Article 7);
- Non-compliance (Article 8); and
- Technical assistance (Article 10), and various other subjects.

The treaty regulates the substances specified in Annexes:

- A-(CFCs, halons);
- B-(other fully halogenated CFCs, carbon tetrachloride, methyl chloroform);
- C-(HCFCs), E (methyl bromide); and
- F-(HFCs).

The treaty undergoes modifications and adaptations in response to emerging scientific, technical, and economic advancements. The Meeting of the Parties serves as the governing body for the treaty, receiving technical assistance from an Open-ended Working Group. Both entities convene annually. The Ozone Secretariat, located at the headquarters of the UN Environment Program in Nairobi, Kenya, provides assistance to the Parties.

5.1.1. The Multilateral Fund

The establishment of the Multilateral Fund for the Implementation of the Montreal Protocol took place in 1991, as mandated by Article 10 of the treaty. The primary goal of the Fund is to offer financial and technical support to developing nations that are party to the Montreal Protocol. These countries have an annual per capita consumption and production of ODS that is below 0.3 kg, and the aim is to help them adhere to the control measures outlined in the Protocol. The activities of the Multilateral Fund are carried out by four international agencies, namely:

- The UN Environment Programme (UNEP);
- UN Development Programme (UNDP);
- UN Industrial Development Organisation (UNIDO); and
- The World Bank.

Additionally, bilateral agencies of non-Article 5 countries also play a role in the implementation of these activities.

The operation of the Fund is overseen by the Executive Committee, which consists of seven members from both Article 5 countries and non-Article 5 countries. The Committee receives valuable assistance from the Multilateral Fund Secretariat, which is located in Montreal (Velders et al., 2007). Throughout its existence, the Multilateral Fund has provided backing for more than 8,600 projects, encompassing a wide range of initiatives such as:

- Industrial conversion;
- Technical assistance;
- Training, and
- Capacity building.

The total value of these endeavors exceeds an impressive US$3.9 billion.

During the implementation of the Montreal Protocol, it has been observed that developing countries have shown their willingness, readiness, and ability to be equal partners in global endeavors aimed at safeguarding the environment, provided they receive appropriate assistance. It is worth noting that numerous developing countries have successfully surpassed the set reduction targets for phasing out ODS, thanks to the invaluable support provided by the Multilateral Fund (Velders et al., 2007).

5.1.2. Phase out of HCFCs – The Montreal Amendment

The phase out of HCFCs, as outlined in the Montreal Amendment, is a crucial step in addressing the environmental impact of these harmful substances. Hydrochlorofluorocarbons (HCFCs) are gases that have been widely utilized across the globe in various applications such as refrigeration, air-conditioning, and foam production. However, it is crucial to note that these substances are currently being phased out under the Montreal Protocol due to their detrimental impact on the ozone layer. HCFCs have the dual impact of being ODS and potent greenhouse gases. In fact, the most widely used HCFC is approximately 2,000 times more powerful than carbon dioxide when it comes to its global warming potential (GWP) (Mäder et al., 2010).

In September 2007, the Parties made the decision to expedite their timeline for phasing out HCFCs, acknowledging the significant advantages this would bring to the Earth's climate. Developed countries have made significant progress in reducing their consumption of HCFCs and are on track to completely phase them out by 2020. Developing countries made the decision to initiate their phase-out process in 2013 and are currently implementing a gradual reduction plan that will lead to the complete elimination of HCFCs by 2030.

The HCFC phase-out is currently underway in Article 5 countries, receiving substantial support from the Multilateral Fund. This support includes:

- The implementation of multi-stage HCFC Phase out Management Plans (HPMPs);
- Investment projects; and
- Capacity building activities.

During this process, it is highly emphasized that all countries should actively promote the adoption of alternatives to HCFCs that have minimal environmental impact, especially in terms of climate change. Additionally, these alternatives should also prioritize health, safety, and economic factors.

When considering the climate, it is important to take into account factors such as global-warming potential, energy use, and other relevant considerations. In the realm of refrigeration and air conditioning, it is imperative to maximize the efficiency of refrigerants, equipment, servicing methods, as well as the processes of recovery, recycling, and disposal at the end of their lifespan (Velders et al., 2007).

5.1.3 The Kigali Amendment: Phasing down HFCs

HFCs were introduced as non-ozone depleting alternatives to facilitate the timely phase-out of CFCs and HCFCs. HFCs have become increasingly prevalent in a wide range of products, including air conditioners, refrigerators, aerosols, foams, and more. Although the stratospheric ozone layer is not affected by these chemicals, it is worth noting that some of them possess significant GWPs that range from 12–14,000. The growth rate of HFC emissions is a concerning 8% per year, and if this trend continues, annual emissions are expected to make up a significant 7%–19% of global CO_2 emissions by 2050. The uncontrolled growth in HFC emissions poses a significant challenge to the global efforts aimed at limiting the rise in temperature to 2°C or below this century. Immediate measures must be taken to address the issue of HFCs to safeguard the integrity of the climate system (Velders et al., 2007).

An agreement was reached by the Parties to the Montreal Protocol during their 28th Meeting of the Parties in Kigali, Rwanda on 15 October 2016. The agreement aims to gradually reduce the use of HFCs. It has been unanimously decided by nations to include HFCs in the list of regulated substances and have also endorsed a specific timeline for their phased reduction by 80%–85%, expected to be achieved by the late 2040s. In 2019, developed countries are anticipated to make their initial reductions. In the coming years, developing countries are expected to implement a freeze on their HFC consumption levels. This freeze is projected to take effect in 2024 for most nations, with some countries following suit in 2028.

This particular matter has been the subject of negotiations between the Parties for quite some time now, dating back to 2009. The recent achievement of the Kigali Amendment (Decision XXVIII/1 and accompanying Decision XXVIII/2) serves as a significant continuation of the remarkable legacy established by the Montreal Protocol. Effective from 1 January 2019, the Kigali Amendment will come into force for the countries that have ratified it.

In order to successfully implement the HFC phase down, it is crucial to decrease reliance on high-GWP alternatives and promote the use of low-GWP, energy-efficient technologies. This should be done as a part of the HCFC phase-out process under the Montreal Protocol. This particular approach has the potential to successfully accomplish the objective of the Montreal Protocol, which is to eliminate HCFCs. Additionally, it can also lead to energy efficiency gains and a reduction in CO_2 emissions, which can be seen as a positive outcome in terms of climate impact (Velders et al., 2007).

5.1.4. The Progress Made so Far and the Tasks that Lie Ahead

Through the diligent and unwavering enforcement of the Montreal Protocol, it is anticipated that the ozone layer will experience a complete restoration by the midpoint of the current century. It is crucial to acknowledge that the absence of this treaty would have led to a staggering tenfold rise in ozone depletion by 2050, in stark contrast to the present situation. This alarming scenario would have had dire consequences, including:

- Millions of additional cases of melanoma;
- Other forms of cancer; and
- Eye cataracts.

One notable estimation suggests that the Montreal Protocol will have a significant impact on public health, potentially saving up to two million lives annually by the year 2030, specifically by reducing the incidence of skin cancer (Mäder et al., 2010).

As of now, the Parties to the Protocol have successfully eliminated 98% of ODS worldwide in comparison to the levels recorded in 1990.

The Montreal Protocol plays a crucial role in safeguarding the global climate system due to the potent greenhouse gases present in these substances. Between

1990 and 2010, the control measures of the treaty resulted in a significant reduction of greenhouse gas emissions. It is estimated that these measures led to a reduction of approximately 135 gigatons of CO_2, which is equivalent to an annual reduction of 11 gigatons.

The Kigali Amendment is set to have a significant impact on reducing greenhouse gas emissions. By limiting the use of HFCs under the Montreal Protocol, it is estimated that up to 105 billion tonnes of carbon dioxide equivalent of greenhouse gases will be prevented. This will help to avoid a temperature rise of up to 0.5°C by 2100. It is worth noting that this is an unprecedented contribution to climate mitigation efforts and the largest effort made by the world to keep global temperature rise "well below" 2°C, as agreed upon at the Paris climate conference (Mäder et al., 2010).

5.2. THE UNITED NATIONS FRAMEWORK CONVENTION ON CLIMATE CHANGE

From its establishment in 1992 until the signing of landmark agreements like the Kyoto Protocol in 1997 and the Paris Agreement in 2015, international climate negotiations have been based on the United Nations Framework Convention on Climate Change (UNFCCC), a foundational treaty. A total of 197 states have pledged to combat climate change and keep the international community apprised of their efforts by ratifying the Convention (Hickmann et al., 2021).

The Convention's main objective is to achieve the stabilization of greenhouse gas concentrations in the atmosphere. This is crucial in order to prevent any harmful interference with the climate system caused by human activities. The Convention aims to accomplish this within a timeframe that allows for both societal and environmental adaptation, while also promoting sustainable economic development. By becoming part of the Convention, Parties recognize the presence of the climate change threat and commit to taking action to address it.

The Convention itself has not set specific targets. Instead, its purpose was to establish a structure for upcoming agreements and policies. The Convention emphasizes the importance of continuous research, negotiations, agreements, and obligations through several key mechanisms as discussed below.

5.2.1. Common but Differentiated Responsibility

This is an important concept that recognizes the need for shared accountability while acknowledging the varying levels of responsibility among different actors. The Convention acknowledges the concept of common but differentiated responsibility (CBDR) and respective capability. This principle recognizes the varying capabilities and responsibilities of different states in addressing climate change. It provides a framework for finding a balance between the need for all states to take action on climate change, considering their different levels of responsibility for greenhouse gas emissions and their capacity to address the problem fairly and suitably (Bakhtiari, 2018).

In light of different national situations, the Paris Agreement recognizes that different countries have different responsibilities and capacities. It takes into account the fact that emissions are dynamic and can change over time, for example, that China's emissions have risen dramatically since 1992 and that our capacity to combat climate change can change over time as well, due to changes in economies, technology, and policy (Hickmann et al., 2021).

5.2.2. Putting in Place Important Organizations, Frameworks, and Procedures

The Convention establishes the framework for the Parties to collaborate in order to accomplish its objectives. When it comes to reviewing, developing, agreeing upon, and implementing climate policy, the highest authority is the Conference of the Parties. A secretariat assists developing nations in particular by coordinating the Conference of the Parties sessions, preparing reports, and compiling information.

In addition to evaluating the Convention's implementation, subsidiary bodies offer recommendations based on science and technology. The Kyoto Protocol and the Paris Agreement, which came later under the UNFCCC, created more organizations like the Green Climate Fund, specialized funds for climate change, and technical committees. The Convention, the Kyoto Protocol, and the Paris Agreement rely on the technical, advisory, and financial advice provided by these institutions to inform policymaking and the development of guidelines (Bakhtiari, 2018).

5.2.3. Information Gathering, Education, and Effective Expression

The UNFCCC process places a strong emphasis on knowledge-building. To establish evidence, establish goals, and monitor progress toward reducing greenhouse gas emissions, all parties involved must collect and share relevant data. Also, developed nations need to lay out their plans for mitigating climate change and the results they anticipate. This procedure has been essential in establishing a solid foundation of information and scientific comprehension regarding climate change and the effects of different mitigation techniques. This foundation has served to guide following legislation and major international accords like the Paris Agreement. The Convention mandates that nations encourage public climate education, awareness, and training because public comprehension and engagement are crucial to attaining national and international transformation.

5.2.4. Ways in Which UNFCCC has been Effective

Several agreements have been adopted since the UNFCCC was established in 1992. Measurement, tracking, and reporting of emissions and impacts; generation of knowledge and research; and building the capacity to address the causes and effects of climate change are all pillars of today's climate action that have been spurred by it. These pillars have been developed at the national and international levels. The adoption of the Paris Agreement, which was produced at the 21st session of the Conference of the Parties (COP21) in 2015, is the most recent significant development in the UN climate change agenda. The Glasgow Climate Pact, which tries to "turn the 2020s into a decade of climate action and support," was adopted by nations in 2021 (Hickmann et al., 2021).

5.3. KYOTO PROTOCOL

The Kyoto Protocol was an international pact that aimed to lower atmospheric concentrations of greenhouse gases (GHG) and reduce emissions of carbon dioxide. A requirement for industrialized nations to reduce their emissions of carbon dioxide was a cornerstone of the Kyoto Protocol. The urgent issue of greenhouse gas emissions and their influence on climate stability led to the 1997 establishment of the protocol in Kyoto, Japan. It was largely superseded by the Paris Agreement when it went into force in 2016 (Pickering et al., 2018).

During a critical period of escalating global warming concerns, the Kyoto Protocol imposed the responsibility on developed nations to reduce their greenhouse gas emissions. The Protocol was associated with the United Nations Framework Convention on Climate Change.

Nations that ratified the Kyoto Protocol were allocated maximum carbon emission levels for designated time frames and engaged in the trading of carbon credits. If a nation were to exceed its allocated limit, it would face penalties in the form of a reduced emissions cap for the subsequent period.

Industrialized nations committed to reducing their annual hydrocarbon emissions by an average of 5.2% by 2012, as part of the Kyoto Protocol. The number of targets varied depending on each specific country. Consequently, every country was assigned a distinct goal to achieve by that particular year. The European Union (EU) committed to an 8% reduction in emissions, while the United States and Canada made promises to decrease their emissions by 7% and 6% respectively, by 2012 (Pickering et al., 2018).

5.3.1. Mechanisms of the Kyoto Protocol

There are three distinct ways that nations can achieve their emissions targets that were set out in the Kyoto Protocol. These three processes were:

1. **Through the International Emissions Trading Mechanism:** *N*ations that had spare emission units could sell them to those that were over-targeted through carbon trading.
2. **Through the Clean Development Mechanism:** *D*eveloped nations that have pledged to reduce or limit their emissions could help poor nations with emission reduction projects so that they could receive credits for their efforts.
3. **By Working Together:** Countries that have pledged to reduce or limit their emissions can earn units that can be used toward that goal through the joint implementation mechanism.

5.3.2. Did Kyoto Achieve Its Goals?

For the first commitment period, the emissions reductions were attained. For instance, compared to 1990, 15 EU nations—including Switzerland—that had pledged to reduce emissions by an average of 8% succeeded in reducing emissions by 11.7%. Nonetheless, the trend indicates that during the same time period, emerging nations like China, Mexico, Brazil, and India as well as developed nations like the USA increased their CO_2 emissions gradually. In comparison to 1990, there was an approximate 29% increase in global greenhouse gas emissions by 2010 (Pickering et al., 2018).

5.3.3. Comparing the Responsibilities of Developed and developing Nations

Greenhouse gas emissions in the atmosphere have reached dangerous levels, and the Kyoto Protocol clearly states that industrialized nations are largely to blame. These emissions have been building up for almost 150 years due to human industry. Consequently, industrialized nations were held more accountable by the protocol than less developed nations.

The European Union and 37 other industrialized nations were obligated to decrease their emissions of greenhouse gases under the Kyoto Protocol. With the request for voluntary compliance from nine developing nations, the Kyoto Protocol was entirely waived for more than a 100 developing nations, including China and India.

Developed nations were categorized as Annex I, while developing nations were categorized as Non-Annex I, according to the protocol. Only nations listed in Annex I were required to comply with the protocol's emission restrictions. There were concerted efforts to lower emissions in each country that was not an Annex I state.

Participating in these projects gave developing nations the chance to earn carbon credits that they could resell or trade with more wealthy nations. The developed nations were able to have a higher maximum carbon emissions level for that specific period thanks to this arrangement. Developed nations were able to keep their greenhouse gas emissions high in large part because of this function.

Despite ratifying the Kyoto Protocol at the outset, the US withdrew its support for it in 2001. The United States believed the deal was unfair because it targeted industrialized countries alone for emissions reduction targets, which it said would hurt American businesses and consumers (Stern, 2018).

5.3.4. Added Changes to the Kyoto Agreement

Around the same time that the Kyoto Protocol became binding international law in 2005, worldwide emissions persisted in rising. It should be noted that the world's emissions increased significantly, by around 40%, between 1990 and 2009. The European Union exceeded its original goal and has stated its confidence in its capacity to accomplish additional reductions in emissions in the future. While other countries were able to meet their targets, the United States and China, as two of the world's leading emitters, produced enough greenhouse gases to cancel each other out (Stern, 2018).

5.3.5. Doha Amendment extended Kyoto to 2020.

Representatives from the Kyoto Protocol met in Doha, Qatar, in December 2012 to authorize an amendment to the original Kyoto agreement, after the initial commitment period of the Protocol had ended. As part of the second commitment period, which began in 2012 and ended in 2020, the Doha Amendment set further reduction objectives for emissions for the participating nations. This meeting was all about climate change as it pertained to the UN.

This Doha Amendment was only in effect for a short time. The Paris Climate Agreement, which essentially succeeded the Kyoto Protocol, was signed by all UNFCCC participants at the 2015 Paris sustainable development summit (Pickering et al., 2018).

5.4. THE PARIS AGREEMENT

Addressing the pressing problem of climate change, the Paris Agreement is a treaty with substantial legal force and universal recognition. On 12 December 2015, at the United Nations Climate Change Conference (COP21) in Paris, France, 196 Parties approved the agreement. On November 4, 2016, the implementation took effect. The main goal is to keep the world's average temperature well below 2°C above pre-industrial levels, with supplementary efforts aiming to keep the increase to 1.5°C above pre-industrial levels. But, in recent years, world leaders have stressed the critical need to limit global warming to 1.5°C by the end of this century. This is since the United Nations' Intergovernmental Panel on Climate Change has warned that the severity of climate change impacts could increase if global warming exceeds the 1.5°C threshold (Delbeke et al., 2019).

The frequency and severity of droughts, heatwaves, and rainfall may increase. To achieve the goal of limiting global warming to below 1.5°C, it is of utmost importance that greenhouse gas emissions reach their highest point by 2025 and subsequently decrease by 43% by 2030. As far as international efforts to combat climate change are concerned, the Paris Agreement is a watershed moment. The treaty binds all signatory nations together in the fight against and adaptation to climate change.

5.4.1. The Significance of Keeping Global Warming to 1.5°C

According to scientists, even a slight increase in temperature, as small as 0.1°C, can have significant consequences for the planet. These include:

- Longer heatwaves;
- More intense storms; and
- An increased risk of wildfires.

It was agreed that the 1.5°C target is crucial due to the overwhelming evidence indicating that the consequences would significantly worsen as the world approaches 2°C. Certain alterations have the potential to become permanent. While the scientific community has not reached a definitive conclusion, it is important to consider the potential outcomes of a 2°C global warming scenario compared to a 1.5°C scenario which include:

- On average, extremely hot days in mid-latitudes would be 4°C warmer, compared to 3°C at 1.5°C;
- At a temperature increase of 1.5°C, the rise in sea level would be 0.1m higher, resulting in a greater number of people, up to 10 million, being exposed to more frequent flooding; and
- The loss of coral reefs would be devastating, with more than 99% at risk of disappearing. This is a significant increase compared to the 70–90% that would be affected at 1.5°C.

5.4.2. The Mechanism of the Paris Agreement

To effectively implement the Paris Agreement, it is crucial that we undergo a significant societal and economic transformation that is firmly based on the latest scientific research. As part of the Paris Agreement, countries are obligated to enhance their efforts in addressing climate change every five years. Nations are required to submit their NDCs, which are plans to combat climate change, by 2020. Each new NDC aims to demonstrate its increased ambition compared to its predecessor.

The COP27 cover decision recognizes the urgent need to act quickly to keep global warming below 1.5°C. As a result, it asks that Parties review and enhance their 2030 targets in their NDCs so that they are in line with the temperature goal of the Paris Agreement by the end of 2023, considering various national circumstances (Delbeke et al., 2019).

5.4.3. Nationally Determined Contributions (NDCs)

Countries outline the actions they will take in their NDCs to reduce greenhouse gas emissions and achieve the objectives of the Paris Agreement. Countries also convey in their NDCs the actions they will undertake to enhance their resilience and adapt to the effects of climate change.

In order to provide a clear direction for achieving the ultimate objective, the Paris Agreement calls upon nations to develop and present long-term low greenhouse gas emission development strategies (LT-LEDS).

LT-LEDS offer a comprehensive perspective on the NDCs for the long term. While NDCs are not mandatory, these, on the other hand, are not obligatory. However, the NDCs are positioned within the framework of long-term planning and development priorities, offering a clear vision and guidance for future development of countries (Savaresi, 2016).

5.4.4. What Actions have Countries Taken following the Paris Agreement?

International summits, called COPs (Conference of the Parties), serve as a platform for world leaders to convene and deliberate on their climate commitments. Since 2015, all COPs have been monitoring the progress made by countries in implementing their commitments made in Paris. Upon the signing of the agreement, it was acknowledged by governments that the Paris targets would not be sufficient in curbing global warming to a level of 1.5°C.

Based on the analysis of the independent Climate Action Tracker, the current pledges and targets indicate that the world is projected to experience approximately 2.1°C of warming by 2100. The UK is obligated by law to decrease emissions by 68% by 2030, in comparison to the levels recorded in 1990. However, in September 2023, Prime Minister Rishi Sunak made the decision to postpone the implementation of various environmentally friendly initiatives, including the delay of the phase-out of new petrol and diesel cars. The independent climate advisers of the UK government have expressed their concern regarding the announcements, stating that they may lead the country further astray from fulfilling its legal obligations. During the latest COP28 conference in December 2023, nations reached a significant milestone by collectively acknowledging the need to shift away from fossil fuels. While it remains voluntary for countries to take action, this agreement marks a crucial step towards a sustainable future. The analysis conducted by the UN for COP28 emphasized the urgent need for significant and continuous reductions in greenhouse gas emissions to avoid the irreversible breach of the 1.5°C threshold.

5.4.5. Measures that Countries are Taking to Provide Mutual Support

Needy nations can access financial aid, technological support, and capacity building opportunities through the Paris Agreement's all-encompassing framework such as:

1. **Financial Aid:** Developed nations must spearhead financial aid to less fortunate and more vulnerable nations, according to the Paris Agreement. Furthermore, it is the first document to bring up the idea of other Parties' voluntary contributions. To successfully reduce emissions, mitigation efforts require climate finance, which is essential because large investments are required. A significant amount of funding is needed to address the difficulties and lessen the impact of a changing climate, which is why climate finance is so important for adaptation.
2. **The Technological Domain:** To reduce emissions of greenhouse gases and increase resilience to climate change, the Paris Agreement emphasizes the significance of completing technology transfer and development. It establishes a technical foundation that provides thorough instructions to the effective Technology Mechanism. Through its policy and implementation branches, the mechanism is propelling technological advancement and dissemination (Delbeke et al., 2019).
3. **Capacity Building:** A lot of poor nations don't have what it takes to deal with all the problems that climate change has brought about. Therefore, developing nations and their capacity building concerning climate issues is a major focus of the Paris Agreement. Additionally, it urges wealthy nations to boost funding for programs that help poor nations enhance their capacity (Savaresi, 2016).

CHAPTER 6

CLIMATE SECURITY AND SUSTAINABLE DEVELOPMENT

Contents

6.1. THE SUSTAINABLE DEVELOPMENT GOALS

The urgent need to put climate promises into action with increasing ambition is underscored by the fact that we are rapidly approaching a tipping point for the increase of global temperatures. Increased financial flows are essential if global governments are to realize their goals of achieving net-zero emissions by 2050.

Governments worldwide should direct investment towards climate and sustainable development impact through globally recognized standard, in response to the increasing demand for sustainable investment and the convergence of voluntary and compliance carbon markets.

To achieve national climate goals through carbon markets and other financial mechanisms, to certify climate impacts under Article 6 of the Paris Agreement, and to contribute to the UN Sustainable Development Goals, governments should follow strict measures for the global goals (UNDP, 2024; Figure 6.1).

Figure 6.1. *The Synergies between climate change, green bonds, and SDGs.*

Source: Climate Bonds Initiative, 2018.

6.2. SUSTAINABLE DEVELOPMENT GOAL 12

Key to maintaining the standard of living for both present and future generations is achieving sustainable consumption and production patterns, the focus of Goal 12.

There are more people living on Earth than ever before, yet our planet's resources are dwindling. The natural resources needed to sustain current lifestyles would require the equivalent of nearly three planets if the global population reaches 9.8 billion by 2050. One of the most important things we can do to lower our consumption is to switch to more sustainable energy sources, which will require us to alter our current consumption patterns. But in 2020 and 2021, fossil fuel subsidies almost doubled due to worldwide crises (Munasinghe, 2011). There have been encouraging shifts in several sectors, such as the growing popularity of sustainability reporting—the number of sustainability reports published has nearly tripled in the past three years alone—which bodes well for the future of the corporate world.

Fighting food loss is an immediate priority that calls for data-driven policies, investments in technology, infrastructure, education, and monitoring, and is another indicator of overconsumption. Even though a large portion of the world's population goes hungry every day, an astounding 931 million tons of food is wasted (Matthew and Hammill, 2009).

6.2.1. We Must Change Our Consumption Habits.

The significant advancements in economic and social spheres throughout the past century have unfortunately been accompanied by a concerning decline in our environment. This degradation poses a serious threat to the very systems that are crucial for our future development and overall survival. To achieve a successful transition, it is imperative to focus on:

- Enhancing resource efficiency;
- Taking into account the complete life cycle of economic activities; and
- Actively participating in multilateral environmental agreements.

6.2.2. What Needs Changing

There are numerous facets of consumption that, through minor adjustments, can yield significant repercussions for society at large. It is imperative for governments to take decisive action by implementing and enforcing policies and regulations. These measures should encompass setting clear targets to:

- Reduce waste generation;
- Promoting the adoption of circular economy practices; and
- Providing support for sustainable procurement policies.

Transitioning to a circular economy requires careful consideration of product design, with a focus on longevity, repairability, and recyclability. One important aspect to consider is the promotion of practices like:

- Reusing;
- Refurbishing, and
- Recycling products in order to minimize waste and resource depletion.

It is important for individuals to consider adopting more sustainable lifestyles. This includes:

- Consuming less;

- Making conscious choices to use products with lower environmental impacts; and
- Actively working to reduce the carbon footprint of their day-to-day activities.

6.2.3. Several Ways that a Business can Offer Assistance

It is imperative for businesses to actively seek out innovative solutions that promote sustainable consumption and production patterns. It is crucial to gain a deeper comprehension of the environmental and social consequences associated with products and services. This includes examining the entire life cycle of a product and how it is influenced by its usage within different lifestyles (Munasinghe, 2011).

Innovation and design solutions have the potential to empower and motivate individuals to adopt more sustainable lifestyles, thereby minimizing their environmental footprint and enhancing their overall well-being.

6.2.4. Strategies that Consumers can Employ to have a Beneficial Influence

There are two primary methods to provide assistance:

- It is important to consider ways to minimize waste; and
- Make conscious choices when purchasing items, opting for sustainable options whenever feasible.

You must refrain from wasting food and make a conscious effort to decrease your usage of plastic, as it is a significant contributor to the pollution of our oceans. To make a positive impact on the environment, it is advisable to consistently employ certain practices. These include:

- Utilizing a reusable bag;
- Abstaining from the use of plastic straws; and
- Ensuring the proper recycling of plastic bottles.

By incorporating these actions into your daily routine, you can contribute to the greater cause of environmental sustainability. It is crucial to make well-informed purchases. By purchasing products from sustainable and local sources, individuals have the power to make a significant impact and encourage businesses to embrace sustainable practices (Matthew and Hammill, 2009).

6.2.5. Target of Goal 12

- To promote sustainable consumption and production, it is crucial for all countries to actively participate in implementing the 10-year framework of programs. Developed countries, being more advanced in this area, should take the lead and guide the efforts, while also considering the specific needs and capabilities of developing countries.

- By the year 2030, it is imperative to accomplish the goal of attaining sustainable management and efficient utilization of natural resources.
- To achieve the Sustainable Development Goals, it is crucial to address the issue of food waste and losses. By 2030, the aim is to reduce per capita global food waste at the retail and consumer levels by half, as well as minimize food losses throughout the production and supply chains, including post-harvest losses.
- Additionally, by 2020, the goal is to ensure the environmentally sound management of chemicals and all wastes throughout their life cycle. This includes reducing their release to air, water, and soil to minimize their negative effects on human health and the environment.
- By 2030, waste generation should be significantly reduced through various measures such as prevention, reduction, recycling, and reuse.
- It is highly recommended that companies, particularly those of significant size and international reach, embrace sustainable practices and incorporate sustainability information into their reporting cycle.
- It is crucial to emphasize the importance of promoting sustainable public procurement practices that align with national policies and priorities. This will help ensure that our procurement processes contribute to long-term sustainability and support our overall goals as a nation.
- To achieve sustainable development and lifestyles in harmony with nature, it is crucial to ensure that people worldwide have access to the necessary information and awareness.
- Additionally, supporting developing countries in enhancing their scientific and technological capacity is essential for them to transition towards more sustainable patterns of consumption and production.
- Furthermore, it is important to develop and implement tools to monitor the impacts of sustainable development, particularly in the context of tourism, which can create employment opportunities and promote local culture and products.
- It is imperative to address the issue of inefficient fossil-fuel subsidies that promote wasteful consumption. These subsidies create market distortions that need to be rectified. One way to achieve this is by

restructuring taxation and gradually phasing out these harmful subsidies. It is crucial to consider the environmental impacts of these subsidies and take into account the unique circumstances of developing countries. It is important to minimize any potential negative effects on their development while ensuring the protection of the poor and affected communities.

6.3. CLIMATE SECURITY AND POVERTY ALLEVIATION

It is becoming more and more evident that there is a strong connection between climate change and the welfare of individuals living in poverty. For both climate action and poverty reduction to be successful, they must be pursued simultaneously. It is important to address the persistent misconception that implementing effective strategies to combat and adjust to climate change is contradictory to efforts to alleviate poverty. Such delays in climate action not only hinder progress but also create confusion in decision-making processes (Munasinghe, 2011). This, in turn, results in fragmented and less impactful climate and development strategies that ultimately have negative consequences for individuals living in poverty.

There are two important aspects to consider when examining the relationship between climate and poverty are:

- The first is the impact of climate change on poverty; and
- The second is the economic effect of climate action on poverty.

When it comes to understanding poverty, it is important to take a comprehensive approach that goes beyond just income. We consider various dimensions of well-being, including education, health, housing, and personal security. From this perspective, it is important to note that economic development should not be solely measured by GDP growth.

It is evident that climate change exacerbates poverty for the majority of individuals. Extensive evidence clearly demonstrates that individuals living in impoverished circumstances, particularly those residing in precarious and susceptible environments, are already disproportionately affected by the consequences of climate change (Matthew and Hammill, 2009). There are several concerning factors to consider are:

- A rise in malnutrition, as well as increased losses from natural disasters impacting various aspects of wellbeing;
- Reductions in income from crop and livestock production, along with work disruptions;

- Uninsured medical costs and high, unpredictable food prices further compound the situation; and
- The potential consequences of conflict.

The impacts of future climate change are highly unpredictable and frequently underestimated by economic models, which in turn poses even more significant threats.

As evidenced by studies, the alarming projections suggest that the impact of climate change has the potential to significantly exacerbate global poverty levels by 2030. The estimated range of 32 million–132 million additional individuals facing poverty highlights the urgency of addressing this issue in order to mitigate its consequences (UNDP, 2024).

Climate change affects individuals in various ways. The intersection of marginalized identities and poverty can significantly increase vulnerability to the impacts of climate change. It is evident that climate change has a disproportionate impact on women and girls (Figure 6.2). This is primarily due to:

- Their limited access to resources;
- Their overrepresentation in vulnerable sectors like agriculture; and
- The added responsibility they bear in ensuring their basic needs are met, such as access to food and water.

Figure 6.2. *Climate change poses more risks for women and children.*

Source: By Gillfoto – Own work, CC BY-SA 4.0, https://commons.wikimedia.org/w/index.php?curid=86178010.

Although the advantages of addressing climate change are clear, there are those who raise concerns about the potential economic impact on impoverished communities. It has been argued by some that transitioning away from fossil energy sources is both inefficient and costly. However, it is important to recognize that in our complex world, there are numerous instances of market failures and inefficiencies. It is crucial to take proactive measures to tackle these issues, as doing so can have positive impacts on both the environment and economic growth.

In addition, it is highly improbable that the future will see the same historical correlations between higher carbon emissions and improved well-being. Due to the significant decrease in costs of low-carbon technology, fossil fuels are no longer the most economically viable choice for increasing energy access or creating jobs (Munasinghe, 2011).

Multiple estimates highlight the significant benefits of taking action on climate change for employment and economic output. There are various opportunities to consider, such as:

- Enhancing material efficiency;
- Making technological advancements;
- Improving networks (cities, digital, and transport); and
- Reducing inequality.

Recent research has highlighted the potential benefits of countries pursuing these avenues in relation to the climate-poverty relationship.

The field of economics has frequently failed to fully acknowledge the urgency and advantages associated with transitioning to a low-carbon, resilient, and equitable model of development. Existing models face significant challenges when it comes to incorporating the dynamic advantages of innovation, learning, and feedback loops that drive institutional and behavioral change, discovery, and economies of scale. Consequently, these models tend to underestimate the economic advantages of the transition, encompassing output, employment, and overall well-being, such as health benefits from reduced air pollution that are particularly significant for individuals living in poverty (Matthew and Hammill, 2009).

6.3.1. Obstacles to Combating Poverty by Addressing Climate Change

Fair, low-carbon, and climate-resilient development can only be achieved if each country gives serious thought to its own situation and makes strategic decisions accordingly. It is critical that the policy reaction goes beyond carbon pricing alone, despite its usefulness. Rather, a holistic strategy should be implemented, taking into account a variety of tools and methods to efficiently manage changes, disruptions, and associated expenses. On the other hand, a lot still needs to be learned about how policies, investments, and projects all work together to promote good development and macro-level trajectories.

There are four crucial factors that must be considered when examining the relationship between climate action and poverty are:

1. **When it Comes to Resource Extraction and the Phase-Out of Fossil Fuels, It is Crucial to Retire Existing Fossil Fuel Infrastructure:** This is necessary to prevent carbon lock-in and to mitigate non-climate-related issues like air pollution. Nevertheless, it is crucial to acknowledge the significant consequences of the energy transition on the well-being of communities in poverty, particularly due to the uneven distribution of supply chains across regions in EMDEs. This disparity can pose a serious threat to the livelihoods of entire communities. By implementing proactive planning, meticulous strategy, adequate financing, comprehensive social protection, and adaptive support, it is possible to effectively mitigate these impacts. One important aspect to consider is the need to prioritize the construction of low-carbon infrastructure over the phasing out of high-carbon industries. Tools and guidance remain necessary, particularly in emerging markets and developing economies (EMDEs), where obstacles such as informal economies, inadequate social safety nets, and limited state capacity pose significant challenges to achieving a fair transition. Furthermore, it is crucial to note that the emergence of new and expanded value chains in transition minerals presents a significant avenue for development. However, it is imperative that these value chains strictly adhere to the most stringent social standards.
2. **Fiscal Redistribution, Carbon Pricing, and Fossil Fuel Subsidies:** Low-income households could see some relief from carbon tax revenues if those funds were redistributed, but the poorest nations already have a harder time getting their hands on public funds, so international transfers are crucial for fighting poverty. Redistributing budget savings to families can mitigate the effects on low-income individuals, even though cutting subsidies for fossil fuels could cause prices to rise. Given the highly individualized nature of compensation, the limited coverage of existing transfer instruments, and the problems affecting EMDEs mentioned above, implementing redistributive measures can be complex. For reforms to be successful, they need to be politically and culturally appropriate and encourage public involvement. Transfers can become more straightforward,

affordable, and accurate as digitalization progresses.

3. **Green Job Creation:** To tackle multidimensional poverty, the jobs generated by climate action must be of high quality. Having a job does not always help people in low-income countries get out of poverty. Policies should aim to reduce gender-based and other intersectional forms of marginalization, as well as inequalities in the distribution of skills and the geographic dispersion of new jobs and industries, in order to ensure that vulnerable people have access to employment opportunities as an alternative to fossil fuels. Aligning development strategies with investments in human capital through training and education is essential, as is ensuring an adequate mix of skills in manufacturing and services.
4. **Social Inclusion and Local Effects:** Despite the declining prices of low-carbon technologies, there may be obstacles to deployment in low-income countries and communities due to financial constraints, lack of awareness, and exclusionary upfront costs. This calls for targeted innovation and support to address these issues and promote social inclusion. To ensure fair land rights, for instance, or to help vulnerable people and groups have their needs met, policymakers should craft these technologies with their needs and rights in mind. The community's voices can be heard and climate solutions can be genuinely beneficial when programs are co-created with their input.

6.4. CLIMATE SECURITY AND GENDER EQUALITY

Although global warming poses a risk to all humans, its effects are not uniform. When it comes to gender inequality and other structural disparities, the effects of climate change are both long-lasting and exacerbated. Particularly in regions where women are economically dependent on labor that is vulnerable to climate change, such as agricultural and manual labor, this is the case. Like most humanitarian and development crises, the climate crisis disproportionately affects women (UNDP, 2024). This is because of all the various kinds of systemic discrimination against women, including:

- The uneven distribution of power;
- The lack of equal pay for equal work;
- The high rate of domestic violence; and
- The gender gap in educational attainment and occupational opportunities.

As an example, women are underrepresented in agricultural decision-making processes and resources, despite their important role in agricultural production.

This is especially true when it comes to climate change and agriculture. Women make up just 12.6% of landowners, despite accounting for one-third of all female workers in agriculture. As a result of their inability to influence resource allocation decisions, women in the agricultural, forestry, and fishing sectors receive just 10% of overall assistance. The fact that women have less access to information regarding adaptation technologies, cropping patterns, and weather events makes them more vulnerable. Reducing vulnerability and building more resilient households and communities can be achieved by increasing women's access to resources. To top it all off, having women at the table increases the likelihood that climate change policies will be adopted and, by extension, that adaptation and mitigation initiatives will be effective in meeting women's needs.

However, women's unpaid care duties frequently obstruct their ability to participate in decision-making and lead on climate change. Worldwide, women do 3 times as much unpaid care work as men do (more than 75%). Due to the unequal burden placed on women in supporting their families and communities during climate-related disasters, this figure increases.

As a result of climate change and other climate-related stresses, women, and girls may find it harder to enter the workforce or attend school, which forces them to spend more time at home and further entrenches their lack of agency. In addition, women, and girls may be more susceptible to gender-based violence in the wake of climate-related catastrophes; thus, they require equal access to decision-making opportunities and high-quality services that are critical to their safety and recovery.

For all these reasons and more, promoting gender equality and the leadership and empowerment of women is essential if we are to find lasting solutions to the climate crisis. Prioritizing gender equality in climate action is more important than ever before (UNDP, 2024).

6.4.1. How can Gender Equality Aid Climate Action

Gender equality plays a crucial role in advancing climate action. By ensuring equal opportunities and representation for all genders, we can harness a diverse range of perspectives, skills, and experiences to tackle the complex challenges posed by climate change. When women and marginalized genders are empowered and included in decision-making processes, they bring unique insights and priorities that can lead to more effective and sustainable solutions. Additionally, promoting gender equality can help address underlying social and economic inequalities that contribute to climate vulnerability, creating a more just and resilient society. A gender analysis is an important first step in determining how men and women are affected differently by climate change. This is useful for pinpointing the exact areas where gender disparities and inequality exist so that appropriate action can be taken. More effective, evidence-based climate policies and actions that address the needs of women and girls, as well as men and boys, can be developed once we have a better grasp of the relative distribution of resources, opportunities, constraints, and power in any particular context (Munasinghe, 2011). It is imperative that we guarantee gender parity in the allocation of funds to climate action when crafting and enacting new policies. That way, in the event of an emergency, everyone can contribute their knowledge and take the lead.

To counter this, to accomplish the profound changes needed to address the climate catastrophe and guarantee social justice, climate policies should be coordinated and integrated with gender equality goals.

6.4.2. Climate Action Initiatives that Take Gender Equality into Account

It is crucial to prioritize the development of women's skills and knowledge to promote their active participation in shaping climate policies and bridging the gender divide in climate decision-making. As an illustration, Bhutan has diligently trained Gender Focal Points across various ministries and women's organizations. This comprehensive training equips them with the necessary skills to effectively coordinate and execute initiatives related to gender equality and climate change.

Considering the various obstacles and disparities that women encounter in terms of accessing resources, credit, technology, employment, and economic prospects, the implementation of innovative financing methods can empower women to make a more significant impact in addressing climate change. As an illustration, Zimbabwe is in the process of establishing a renewable energy fund to create targeted entrepreneurship opportunities for women. In Uzbekistan, a pilot green mortgage scheme facilitated the provision of affordable low-carbon energy technologies to rural households in 5 regions. Due to the gender-responsive criteria and targeting of the scheme, a significant 67% of mortgages were acquired by households headed by women (UNDP, 2024).

Enhancing the gender-responsiveness of climate finance yields multiple benefits such as:

- Enhancing the efficacy;
- Enhancing the efficiency; and
- Longevity of investments while simultaneously bolstering women's empowerment and resilience.

It is imperative for countries to take responsibility for monitoring their progress in achieving gender equality and evaluating the impact of their climate interventions on both women and men. As an illustration, Uruguay has implemented a monitoring, reporting, and verification system that focuses on gender equality and women's empowerment. This system tracks the progress of their Nationally Determined Contribution (NDC) programming in this regard. In 2021, the system underwent significant improvements and was subsequently relaunched (StudySmarter UK, 2024).

It is imperative for national ministries with a responsibility to promote gender equality to be actively involved in coordinating efforts for NDC implementation and decision-making processes concerning climate impacts and actions on a larger scale. In Cabo Verde, the Institute for Gender Equality and Equity is a prominent agency that takes the lead in defining the specific needs, targets, and indicators associated with gender mainstreaming in climate action across various sectors and intervention areas (Matthew and Hammill, 2009).

6.4.3. Countries Making Progress in Promoting Gender Equality in Climate Action

Some nations are taking steps toward achieving their NDCs by consciously including gender considerations in climate action, including Chile, Georgia, Lebanon, Cambodia, and Uganda. As it has revised and implemented its NDC, Chile has made efforts in recent years to incorporate gender dimensions even more thoroughly. The establishment of a Gender and Climate Change Roundtable, which produced a set of guidelines for how to include gender considerations in climate change tools, was a significant result. Additionally, Chile's Ministry of Finance has been collaborating on a new approach to evaluate climate change spending in public budgets with a gender perspective.

As part of their NDC revisions, Uganda and Lebanon conducted gender analyses to identify gender gaps in the adoption of adaptation and mitigation strategies in critical areas and to better understand the unique ways in which women and men are affected by climate change. One example is the document "How to Put Gender-responsive Climate Solutions into Action: Lebanon's Approach," which serves as a framework for mainstreaming gender in climate action. It was informed by the Lebanon analysis, which offered entry points to make climate change policies, strategies, planning, and reporting more inclusive and responsive to gender (UNDP, 2024).

To prioritize adaptation and mitigation efforts, gender was one of the criteria included in Cambodia's NDC. As a result, the percentage of female participants in NDC priority actions varied between 15% and 70%. Further, the NDC offers indicators to track this shift and proposes gender-responsive approaches that let women have a real say in certain climate measures (Munasinghe, 2011). As part of its NDC, Georgia has produced a handbook for all government employees involved in environmental issues, outlining best practices for integrating gender into climate change interventions. This resource details how climate change affects women and girls more severely than males and suggests gender-responsive actions that important industries like agriculture, forestry, energy, transportation, building, and industry can take to mitigate the effects of this global warming.

6.4.4. Climate Action with Gender Considerations Faces Many Challenges.

Gender equality was only ostentatiously discussed in the first round of NDCs, which began in 2015, with few countries proposing concrete policies and actions. Many nations' efforts to incorporate gender considerations into their climate policies and action plans have been hindered by this. During the development of the NDC, there was a limited understanding and evidence base of how climate impacts vary between women and men due, in part, to the lack of gender-disaggregated information and data. Furthermore, climate action planning isn't necessarily gender-responsive due to the underrepresentation of women's groups and other civil society organizations in the policymaking processes surrounding climate change. If climate action is to be gender-responsive, it must take into account the unique needs and vulnerabilities of women and men, which can only be achieved by consulting and including women's organizations in planning and decision-making processes (Martin, 2022).

It is crucial to assist nations in turning their goals of gender equality into tangible policies and programs, as the first round of NDCs failed to address this issue. Effective governance, inclusive planning, and integrated policy frameworks were the three pillars upon which the UNDP Climate Promise built its multi-pronged strategy to tackle these problems and lay a firmer groundwork for future advancements in gender-responsive climate action. To better understand women's and men's roles in priority NDC sectors, the structural inequities women encounter when it comes to climate action, and the possibilities for empowering women to be a part of climate solutions, the Climate Promise advocated for thorough gender analyses.

From 2015–2020–2021, the quality of NDCs increased significantly thanks to this strategy, which expanded on previous attempts to achieve more systematic gender mainstreaming within the five-year cycle of NDC revisions. Gender is now mentioned in 95% of NDCs, up from 47% in 2015, and gender-responsive adaptation measures are included in 52% of revised NDCs, up from 11% in 2015 (Matthew and Hammill, 2009).

Changes in political, economic, and social climates pose an ongoing threat to advancements in gender equality, notwithstanding these successes. Many policy,

governance, and financial obstacles must still be overcome before NDCs can be implemented in a gender-responsive manner and development gains can be preserved. Building institutional capacities on gender-responsive climate action is essential, as is enhancing coordination mechanisms between ministries. While governments strive to strengthen climate adaptation and mitigation efforts in the coming years, they will face the formidable task of assisting women in leadership roles in changing the systems and power structures that perpetuate gender inequality (StudySmarter UK, 2024).

More people need to know and understand the case for directing climate funds towards climate action that is responsive to gender. As people fight to adapt to the effects of climate change, inequality increases the burden of development costs on society as a whole.

The transformation needed to combat climate change and secure a sustainable and equitable future for all cannot be achieved in the absence of gender equality.

6.5. CLIMATE SECURITY AND EDUCATION

Fundamentally, development is the act of putting money into a country so that it can grow more rapidly in the future. The return on investment (ROI) for investments in human capital and education is among the highest of any investment strategy. Compared to investing in physical capital, the long-term return on investing in human capital can be more than 10 times higher in certain low and middle-income countries (LMICs) (Munasinghe, 2011).

Climate change will impact investments in human capital just as it does in physical capital, and the education sector, along with the rest of the economy, will have to adapt to the new normal of a hotter and more unpredictable climate. The best way for educational institutions to tackle the problems caused by climate change is still unclear.

Reducing emissions "mitigation" and adjusting to the new normal "adaptation" are two sides of the same coin when it comes to combating climate change. At present, the education sector does not contribute much to global emissions, even though all parts of the economy must drastically cut their carbon emissions for a low-emissions future to be possible. Although nationally representative data is scarce, it seems that the education sector accounts for about 2%–3% of a country's emissions in high-income nations, with the majority of this contribution coming from purchased power (Martin, 2022).

According to US estimates, only 2% of emissions come from US universities and colleges; the remaining 70% come from Scope 2 and 3 emissions (Matthew and Hammill, 2009). Similarly, none of South Africa's schools use more than 2% of the country's total energy. In comparison, 24% of U.S. greenhouse gas emissions come from the industrial sector and 11% from agriculture. In contrast, there is room for the education sector to make a significant impact in the adaptation process. There is a large and quickly expanding body of research that shows how climate change, and especially extreme heat, is bad for kids. According to predictions, a warmer planet will subject large portions of the globe to much higher temperatures, making this link crucial.

Additionally, families in agricultural communities are more likely to take their children out of school during years when high temperatures cause larger crop losses. This has a negative impact on their human capital formation.

Reducing emissions from schools alone might not be as effective as addressing the negative effects of heat exposure on human capital formation through encouraging investments in adaptation. Consider this example to gain a deeper understanding. Imagine a significant US high school in the southeast region taking the initiative to install solar panels on its roof, thereby reducing its greenhouse gas emissions from electricity consumption by 50%. When we consider the carbon intensity of electricity in the southeast, the average electricity consumption in schools, and the U.S. government's social cost of carbon, which stands at $50, we can arrive at an estimate of approximately USD 1.5 million for the global benefits that can be achieved by avoiding tons of CO_2 emissions over the next decade. The school could have saved around $2.6 million in future earnings if it had spent that money on air conditioning for all of its classrooms instead of solar panels. However, there might be different costs associated with air conditioning than solar panels (StudySmarter UK, 2024).

That being said, the education sector should not limit its engagement with climate change to the adaptation front alone. In any event, schools might stand to save money on their energy bills—those savings could go toward increasing teacher salaries, for instance. Improved temperature control is just one of the many benefits of installing energy-efficient windows, insulation, or heat pumps in school buildings.

Furthermore, there is hope that the field of education can help shift public perceptions regarding climate change. However, studies on the efficacy of education alone in bringing about social change are contradictory. Research conducted in the United States has shown that individuals who completed a one-year intensive course were able to reduce their self-reported personal emissions by 10%. It is worth noting that a significant portion of this reduction can be attributed to the decision to purchase an alternative vehicle (Munasinghe, 2011).

While there are many ways in which education can fight climate change, one of the most important is preparing schools for the effects of the changing environment. Those with more education appear to be better able to adapt to and weather the effects of climate change. To combat climate change, the education sector must advance adaptation to the changing climate so that it can accomplish its goals of improved learning outcomes.Climate change poses a threat to education, which is a potent growth engine. It is essential that education policy and research find ways to modify classroom practices and school infrastructure to mitigate the adverse effects of climate extremes. Finding and advocating for climate adaptations for young learners could be the most impactful way for the education sector to tackle the climate change challenges, given the huge, compounding effects of enhancing students' learning abilities and human capital accumulation from an early age (Matthew and Hammill, 2009).

6.6. CLIMATE SECURITY AND SUSTAINABLE AGRICULTURE

The concept of food security, which refers to the fair and constant availability of healthy food, is being more and more jeopardized by the effects of climate change. Traditional farming is disrupted by unpredictable weather patterns and extreme events like floods and droughts, which cause crop failures and worsen food insecurity, particularly in vulnerable communities. On the other hand, some farming methods make these problems worse. Global warming and the further instability of weather patterns are fueled by greenhouse gas emissions, which are produced by deforestation, livestock farming, and synthetic fertilizers. The production of food is greatly endangered by this vicious cycle (Martin, 2022).

Sustainable agriculture is a solution that seeks to maintain or increase agricultural productivity while decreasing environmental degradation. Below are some of the fundamental principles to attain sustainable agriculture.

6.6.1. Plants Adapted to the Climates of the Nordic and Baltic Regions in the Future

It is imperative to conduct research on subjects that are crucial for ensuring food security and adapting to climate change. This includes studying the impact of projected climate change on the physiology, productivity, and resilience of plants in the face of various abiotic (such as extreme temperatures, drought, and waterlogging) and biotic (such as pathogens and pests) stressors. Furthermore, it is imperative to conduct thorough research on the various factors that influence plant yield and resistance, such as:

- Genotype;
- Root system;
- Microbiome; and
- Environmental conditions.

Understanding the interactive effects of these factors is crucial in identifying genotypes that can effectively adapt to the challenges posed by climate change.

6.6.2. Heightened Production of Food and Feed Protein on a Local and Regional Level

For the sake of achieving sustainable food and feed security, it is imperative that we prioritize the consumption of locally-grown protein. It is crucial to have cultivars that are well-suited for northern growing conditions, as well as researching agronomic practices to understand their environmental benefits and potential risks for pests and diseases.

6.6.3. Plants and Soil Serve as a Crucial Carbon Sink in our Ecosystem.

Further research is required to determine the most cost- and resource-efficient measures for sequestering and storing carbon in plants (such as cover and perennial crops) and soil, with minimal negative impact on the environment, from a climate change mitigation perspective. Understanding the intricate relationship between plants, soil, and microbial biodiversity is of utmost importance to effectively manage the ecosystem's carbon balance and promote sustainable agriculture in the face of a changing climate (Martin, 2022).

6.6.4. Transition to Climate-Smart, Profitable Local and Regional Agriculture

Transformation towards climate-smart and profitable local and regional agriculture is a crucial goal that we must strive to achieve. It is imperative that we implement sustainable practices and innovative strategies to ensure the long-term viability of our agricultural systems. By adopting climate-smart techniques, we can mitigate the impact of climate change on our crops and livestock, while also maximizing profitability for farmers and communities. This transformation requires a comprehensive approach that encompasses research, education, policy changes, and collaboration among stakeholders. Together, we can create a resilient and prosperous agricultural sector that not only adapts to the challenges of climate change (Munasinghe, 2011).

It is imperative to make adjustments to agricultural policies and incentives to foster a shift towards climate-smart agriculture, while also ensuring the profitability, social sustainability, and security of supply in agriculture.

CHAPTER 7

NON-STATE ACTORS SHAPING CLIMATE GOVERNANCE

Contents

INTRODUCTION

Scholarly works and international policy discussions have recently focused on the function of non-state actors in global governance. More and more, non-state actors like indigenous peoples' groups, trade groups, and international environmental organizations are stepping into roles and responsibilities formerly played by nation-states. There are three main perspectives on this development:

- One view sees it as a positive step towards achieving fair and effective solutions to global problems;
- There are those who view it as a mere diversion; and
- There are those who perceive it as undermining the efforts of the state.

There is a debate surrounding the role of non-state actors in decision-making processes. On one hand, proponents argue that these actors bring additional perspectives to the table. However, critics argue that non-state actors lack defined supporters and often bypass democratic procedures (Kuyper, Linnér, and Schroeder, 2018).

Concerns regarding the legitimacy and authority of global governance are heightened by the emergence of non-state actors. Neoliberalism and globalization have contributed to this trend by highlighting the benefits of private sector involvement in the economy and the efficiency gains that come from it. The long-established model of global governance based on nation-states is under attack from a variety of sources, including the proliferation of global communication networks and the expansion of international trade and investment.

Global climate change governance is one area where non-state actors are clearly involved. The course of climate action is influenced by these players, who take part in various ways and on various levels. A further indication of the role of non-state actors in international climate action is the implementation of initiatives like the Lima-Paris Action Agenda (Nasiritousi, Hjerpe, and Linnér, 2016).

7.1. NON-STATE ACTORS

Conceptualizations of global governance often take into account the significance of non-state actors such as NGOs, firms, and epistemic communities in the governance of global issues.

Furthermore, it is important to note that the majority of approaches view NSAs as internal components of the global governance system, rather than external entities. This position goes beyond state-centred approaches, such as neorealism, and challenges the classical transnationalist view on international politics. According to this view, non-state actors are seen as external entities with limited influence, at best (Bäckstrand et al., 2017). Nevertheless, NSAs are increasingly becoming integral to the formation and structure of global governance networks, both in formal and informal settings.

Non-state actors, also known as NSAs, are a diverse group of actors that operate on the international stage and have the potential to impact international relations. These actors are not representatives of states, but they play a significant role in shaping global affairs. There are five main categories of organizations that play a significant role in global affairs namely:

- Intergovernmental Organizations (IGOs);
- International Non-Governmental Organizations (INGOs);
- Corporate Interest Groups (CIGs) and Transnational Corporations (TNCs);
- Epistemic Communities (ECs); and
- The fifth category includes various other organizations such as terrorist networks, professional organizations, scouts, churches, and more.

4INGOs refer to international non-profit, non-violent pressure groups that aim to achieve specific public objectives and exert influence on international political outcomes. Some notable examples include Greenpeace International, Oxfam International, Amnesty International, WWF, and Pax Christi. One important consideration is whether Corporate Interest Groups should be included in the INGO definition or addressed separately. These organizations, such as the International Chamber of Commerce and the World Business Council for Sustainable Development, are not profit-oriented themselves, although their members are. In addition, it is worth noting that these groups are known for their commitment to non-violence and their focus on achieving public goals, such as economic development. Furthermore, they actively strive to exert their influence on the global stage of international politics. From an ideological and functional standpoint, these pressure groups exhibit distinct differences. Hence, a clear differentiation is drawn between NGOs, also known as Civic Pressure Groups, and BINGOs, which are Business NGOs or Corporate Interest Groups.

According to Kuyper, Linnér, and Schroeder (2018), a TNC is characterized as a significant, profit-driven, commercial entity that operates in multiple countries, with offices and/or production units. Some examples of well-known companies include Shell, BP, Ford, GM, Unilever, McDonalds, IBM, Microsoft, Deloitte, and Touche, and many more. Currently, there are approximately 37,000 transnational corporations operating within the international political economy

(Bäckstrand et al., 2017). In most cases, these companies refrain from direct involvement in politics. However, the non-governmental organizations that advocate for their interests are actively engaged in political activities. In recent times, there has been a noticeable inclination for companies to actively pursue their interests on a global scale. They are now able to assert their influence in international organizations like the EU and WTO.

7.2. CLIMATE CHANGE FRAMEWORK CONVENTION AND NON-STATE SECTOR

The primary objective of the United Nations Framework Convention on Climate Change, which was established in 1992 during the Rio Conference, was to "stabilize atmospheric concentrations of greenhouse gases at a level that would prevent dangerous anthropogenic interference with the climate system. In order to ensure that ecosystems can naturally adapt to climate change, prevent any threats to food production, and promote sustainable economic development, it is imperative that we achieve this level within a specified time frame. That goal cannot be attained without the participation of NSAs, as it necessitates a transition away from GHG production in the primary, secondary, and tertiary sectors of national and subnational economies worldwide (Kuyper, Linnér, and Schroeder, 2018).

States (and the EU) are the only entities with the authority to negotiate, make decisions, and establish additional protocols in support of the Convention because it is an intergovernmental agreement.

Given the crucial role that NSAs play in the climate equation and their absence of official representation at the UNFCCC, two significant ways in which NSAs are involved in global climate governance have emerged.

First, NSAs have played a substantial role in the intergovernmental negotiations at the UNFCCC and before its establishment, even though they are not official Parties to the Convention. Representatives from non-governmental organizations were indeed allowed to take part in the UN General Assembly's Intergovernmental Negotiating Committee negotiations, which gave rise to the UNFCCC. The UNFCCC negotiations were heavily influenced by no less than 77 non-state actors (NSAs) due to their technical knowledge, information sharing, advocacy, and lobbying efforts (Bäckstrand et al., 2017). Even now, at the UNFCCC's annual conferences called the COPs, NSAs continue to engage in this kind of behavior. In accordance with Agenda 21, the UNFCCC Secretariat acknowledges nine types of NSA as constituencies that send observers to the yearly COP1. Parties decide whether or not to admit new NGOs to Observer status at each COP, although NGOs can apply to join at any time. A frequent way for private companies to gain access to the COP is to obtain accreditation through business NGOs that have already been admitted as Observers. This is because private companies cannot directly obtain Observer status. Many non-governmental organizations (NGOs) and other NSAs lack observer status, so this is also true for them.

Secondly, a "regime complex" for climate governance has been in place since the early 2000s, as NSAs have been increasingly involved in governance processes outside the UNFCCC due to limited access and frustration with the official process. Outside of the official process, coalitions of NSAs and a handful of state leaders started making commitments and taking action during this time. In response to the UNFCCC's inaction, "clubs" like the Group of 20 and the Major Economies Forum were formed as a stopgap measure. Transnational climate governance is the term used to describe this type of action on a global scale.

This is considered to have reached a turning point when the Parties failed to establish a concrete, legally enforceable agreement in 2009 at Copenhagen. Some have referred to the dramatic increase in the number of groups working together to implement this regime change as a "groundswell" of climate action or a "Cambrian explosion." Consequently, this non-formal development has been noticed by the Parties, who have acknowledged its importance for carrying out the Convention. As presidents of their respective COPs, the Peruvian and French governments, along with the United Nations secretary-general's office and the UNFCCC secretariat, collaborated on the so-called Lima-Paris Action Agenda from 2014–2015. Its sole purpose was to encourage more international climate action (Kuyper, Linnér, and Schroeder, 2018).

The UNFCCC secretariat created the NAZCA portal, which was the first global repository of its kind. To highlight the scope of the existing action and its potential to advance the UNFCCC goals, this work was introduced as a central feature of the COPs in 2014 and 2015.

In the decisions that accompanied the 2015 Paris Agreement, the Parties established a yearly high-level meeting on NSA climate action to be held during the COP, demonstrating their agreement on the significance of these processes. To encourage more of this kind of activity in the actual economy, the Parties have also designated two global champions to serve on a rotating basis for two years.

According to scholars, the paradigm of post-Paris climate governance is "hybrid multilateralism" due to the merging of informal and formal arenas of governance. Following the same accreditation procedure as mentioned earlier, NSAs are able to attend these high-level annual meetings. Hence, NSA attendance at the COP is likely to skyrocket in the years to come (Bäckstrand et al., 2017).

7.3. NON-STATE ACTORS AND HYBRID MULTILATERALISM: AUTHORITY, LEGITIMACY, AND EFFECTIVENESS

7.3.1. Authority

This is a crucial aspect that should not be taken lightly. It plays a significant role in maintaining order and ensuring that rules and regulations are followed. Without authority, chaos can easily ensue, leading to a breakdown in structure and discipline. It is important to recognize In the current era of hybrid multilateralism, it becomes crucial to examine the location of political authority. This question has been a topic of great interest among climate governance scholars in recent years.

With over 12,000 contributions to the NAZCA database, it becomes evident that non-state actors have become significantly engaged in the governance of climate change. This highlights the emergence of private authority in this domain. While certain entries in the database are directly tied to the regulatory efforts of nation-states, numerous others showcase the independent climate action initiatives that have flourished since COP 15.

According to Nasiritousi, Hjerpe, and Linnér (2016), the Copenhagen conference was a significant moment in global climate governance. It signaled a shift towards greater involvement from cities, NGOs, and corporations in addressing climate change. This was accompanied by the emergence of urban engagement and transnational city networks in climate governance.

There have been several studies conducted that have examined the increase in private emission accounting standards, certification schemes, transnational city networks, and public-private partnerships (Bäckstrand et al., 2017). As cities, corporations, and NGOs have taken the initiative to establish their own rules and standards, they have moved beyond simply adhering to the directives of nation-states or intergovernmental treaties. They have assumed positions of governance and created independent realms of power separate from the central government (Kuyper, Linnér, and Schroeder, 2018).

During the time after the Copenhagen conference, certain scholars saw the transfer of responsibility from nation-states to private actors as an indication that the power and influence of states in global climate governance was diminishing. The apparent failure of nation-states to effectively address climate change has significantly reshaped the political landscape in terms of both authority figures and methods of validation.

The Paris Agreement establishes a complex dynamic between sovereign states and non-state actors, as well as between multilateral and transnational climate action. It challenges the assumption that power and authority are shifting away from states in the face of a more intricate climate regime. According to Bäckstrand et al. (2017), this volume proposes that there is a strong connection between public and private authority in the current international climate cooperation scenario. States have granted authority to private actors, as exemplified by the Clean Development Mechanism of the Kyoto Protocol. Additionally, private actors have the ability to create their own rules and standards.

The presence of transnational climate governance and experimentation has been significant, but the extent of its scale and its connections to intergovernmental regimes have been unexpected. Therefore, instead of focusing on determining where authority lies in a time of hybrid multilateralism, it is imperative to delve into a more detailed empirical analysis of how political authority is exercised within the intricate framework of climate governance (Kuyper, Linnér, and Schroeder, 2018). One interpretation of this is that it represents a resurgence of state authority, as evidenced by the ongoing discussion surrounding the legal standing of the Paris Agreement.

The successful functioning of transnational climate governance relies heavily on the establishment of intergovernmental frameworks by states. The fundamental structure of the Paris Agreement consists of the voluntary NDC pledges made by the states, which are not subject to multilateral negotiations or legal obligations. The agreement signifies a shift in global climate politics, where domestic politics now shape international agreements. There are multiple perspectives through which this can be observed. It is important to note that the state plays a crucial role in reinforcing sovereignty. It is worth mentioning that the Paris Agreement does not include any legally binding provisions when it comes to domestic action (Nasiritousi, Hjerpe, and Linnér, 2016). The legally binding aspect lies solely in the process of monitoring progress in NDCs. Furthermore, it is important to note that the Paris Agreement does not require an evaluation of individual states, but rather focuses on the combined efforts and consolidation of all Nationally Determined Contributions (NDCs).

The Paris Agreement seeks to address the absence of legal obligation by establishing a mandatory transparency framework, which aims to create reputational risk. Furthermore, it is worth noting that although the COP decision provides some information about non-state actors, the Paris Agreement itself does not give them much consideration and does not explicitly assign them a formal role in the periodic review.

Under the framework of the Paris Agreement, there has been a significant shift in authority towards state power, thanks to the bottom-up voluntary pledging approach. Climate negotiations play a crucial role in bringing countries together to coordinate their efforts towards common goals. Therefore, the period following the Paris Agreement has the potential to reinvigorate academic interest in the concept of the green, ecological, or environmental state as a catalyst for significant change towards low carbon societies, from local to national and comparative viewpoints (Kuyper, Linnér, and Schroeder, 2018).

7.3.2. Legitimacy

The legitimacy of non-state actors within the United Nations Framework Convention on Climate Change (UNFCCC) has been a subject of considerable interest ever since the Copenhagen Accord. Within the Paris Agreement, there was a notable level of legitimacy in terms of the involvement and inclusion of both state and non-state actors. The French COP presidency received high acclaim for its exceptional leadership in the realm of conference diplomacy, as well as for its remarkable ability to guide and facilitate an all-encompassing COP 21. This involved extensive consultations with all major countries and negotiation groups, ensuring a truly inclusive and collaborative process. It is important to note that there exists a significant disparity in the level of engagement of developing countries in climate action. The majority of climate action initiatives, as assessed by LPAA and NAZCA, do not effectively address the needs of impoverished and vulnerable populations. The legitimacy of non-state action in terms of throughput is expected to generate increased attention in the post-Paris context. This is particularly relevant considering the existence of platforms such as LPAA, NAZCA, and GCA, which provide opportunities for engagement between state and non-state actors. The Paris Agreement offers various channels through which non-state actors can hold states and other non-state actors accountable for their actions and inactions. The transparency framework and Capacity-building Initiative for Transparency play crucial roles within the Paris Agreement. However, the involvement of non-state actors in the hybrid architecture of the agreement can be somewhat ambiguous. The way in which decisions are made will have a significant impact on the legitimacy of the Paris Agreement's throughput. Deliberative quality revolves around the ability of decision-makers to justify their actions in a manner that is acceptable to all parties involved. The involvement of non-state actors, particularly through the TEP and TEM process, will play a pivotal role in this matter. In the realm of hybrid multilateralism, there is a growing conversation between the UNFCCC and non-state actors regarding the role of non-state contributions in LPAA, NAZCA, and the emerging GCA (Kuyper, Linnér, and Schroeder, 2018).

7.3.3. Effectiveness

The field of transnational climate governance has prompted studies into the best ways for the United Nations to encourage decarbonization and climate resilience, as well as to limit global warming to 2°C or less. A high degree of state engagement and compliance, along with ambitious and rigorous commitments, determines an international agreement's efficacy. By engaging in problem definition, setting agendas, shaping rules and principles, building capacity, mobilizing public engagement, evaluating, and monitoring compliance, and aligning comprehensive frameworks of non-state and intergovernmental action, non-state actors play a crucial role in fostering ambition and encouraging participation.

For climate action to be effective in the post-Paris era, it is necessary to coordinate efforts outside the UNFCCC to reduce emissions. In addition to the NDCs, the total of voluntary commitments and actions by non-state and sub-state actors will help reduce greenhouse gas emissions, advance adaptation and mitigation objectives, and accelerate decarbonization. In contrast to the Kyoto Protocol's requirements for reduced emissions, the Paris Agreement does not impose any legal obligations on states concerning their voluntary commitments.

Cooperation between state and non-state actors could increase effectiveness by increasing compliance and the likelihood that both will be willing to step up their ambitions and participate in the future. The efficacy of openness, global stocktaking, and funding is an important concern. The adoption of the Paris Agreement began a process for increasing the ambition of mitigation efforts, and non-state actors are vital in ensuring that NDCs are being monitored effectively.

A variety of voluntary mechanisms for naming and shaming—or the inverse—within the framework of transparency outside of the UNFCCC will determine the effectiveness, in the end.

7.4. UNDERSTANDING THE IMPORTANCE OF GLOBAL CLIMATE CHANGE GOVERNANCE

An important task for the current century is to find a way to control global warming. According to academics, climate change is a wicked problem due to its complexity and the fact that there are no easy answers to it (Bäckstrand et al., 2017). There are socio-political obstacles that add to the difficulty of tackling climate change, in addition to the scientific uncertainties associated with comprehending complicated climate systems and the environmental impacts of these systems. One of these difficulties is that the burning of fossil fuels and changes in land use are two of the main causes of global warming. These two factors are connected to almost every human activity and are thus essential to modern economic growth models.

Because the impacts of most greenhouse gas emissions on the climate take a very long time to materialize, concerns about equity between generations naturally arise. Everyone, to varied degrees, is susceptible to the effects of climate change since there is no correlation between the source of emissions and their subsequent consequences for the majority of greenhouse gasses (Kuyper, Linnér, and Schroeder, 2018). Climate change is a global problem because it has both immediate and long-term effects on people's lives. Economic competitiveness, agricultural, and forestry policies, and disaster-risk reduction are just a few of the many fundamental human development concerns that climate change affects. Put together, these obstacles make it hard to muster the political will necessary to enact climate policy on a large enough scale to deal with the causes and consequences of climate change.

A multitude of issues necessitate action, and the necessity to engage various stakeholders is being more acknowledged. Climate change governance encompasses all deliberate mechanisms and measures that help to express, decide upon, and implement collective interests related to climate change. The goal is to avoid, mitigate, or adapt to climate change, although it is not always the primary or consistent aim. Since emissions reductions are only one aspect of climate change, this definition encompasses a broader spectrum of concerns related to sustainable development.

Climate change is a complex political issue that touches on many other areas, including energy, finances, food security, and health. As a result, many different groups and individuals are operating across various scales, in different regions, and are seeking to mobilize a wide range of discourses, tools, techniques, and practices for governance (Bäckstrand et al., 2017).

The characteristics that distinguish global climate change governance are the multi-level cooperation it necessitates, the inclusion of various actors, and its transnational scope. Thus, climate change policymaking is a miniature version of the larger field of international environmental regulation.

Every year, 1000s of non-state actors attend the international climate change conferences hosted by the UNFCCC. In addition, there are a great number of non-state actors who do not have accreditation taking part in activities and events outside of the conference site. An estimated 50,000 people, including 25,000 delegates from both official states and non-state organizations, flocked to COP 21 in Paris in December 2015. To shape future climate change policies, non-state actors participate in these and other international conferences.

7.5. GLOBAL ENVIRONMENTAL GOVERNANCE AND NON-STATE PLAYERS IN CLIMATE CHANGE

The role of non-state players in climate change and other areas of global environmental governance has grown in recent years. Scholars disagree, however, on how much sway non-state actors should have in an international system where states have long held sway. The realist view maintains that non-state actors can only act in a way that states permit them to and that this power is a direct reflection of the interests of the states themselves. The sociological view, in contrast, holds that information, accountability, and discourse allow non-state players to impact policy outcomes apart from states. From this vantage point, the dynamic between nation-states and non-state actors is decentralized and free-flowing. Understanding the functions of non-state actors in the governance of climate change requires consideration of both viewpoints. Both the realist and sociological schools of thought center on structures and actors, but the former places more emphasis on ideas and processes. The determination of states' material interests and the shaping of the strategy space for states can be better understood when these viewpoints are combined. Through partnerships with states or efforts to influence state policy via lobbying and advocacy, non-state actors can contribute to the shaping of policy outcomes. Additionally, they can exhibit entrepreneurial spirit by acting autonomously on the ground, establishing transnational initiatives that influence actor behavior in their own right.

Within this more expansive framework of government, ideas like legitimacy and authority acquire fresh significance. The goal of non-state actors, in contrast to states, should be to gain legitimacy and power for governance (Kuyper, Linnér, and Schroeder, 2018).

7.6. CLIMATE CHANGE AND THE ROLE OF NON-STATE ENTITIES IN VIOLENT CRIME

More opportunities for violent nonstate actors (VNSAs) to exert power are likely to arise as a result of climate change. This is because climate impacts can:

- Hinder governance;
- Diminish state capacity and legitimacy
- Heighten competition for resources and livable territory; and
- Require invidious policies.

There are six climate-related variables that contribute to the emergence and growth of VNSAs. These variables encompass:

- De facto states;
- Insurgencies;
- Criminal organizations;
- Private security firms;
- Paramilitaries; and
- Terrorists.

There will be various opportunities that will emerge and the demand for VNSAs will be influenced by six different factors. One of these factors is the decline in state capacity and legitimacy caused by crises related to food, water, and energy. There are several factors that can contribute to an increase in the demand for VNSA activity. These factors include the presence of environmentally inhospitable areas, stricter restrictions on resources, a higher demand for people smugglers and armed border guards, the chaos and injustice that often follow climate-exacerbated disasters, and the growing frustration towards those who are deemed responsible for climate change.

Powerful nonstate actors may find it easier to rationalize their breach of the state's violence monopoly if climate change lowers a state's legitimacy as a result of its performance or outcomes.

The theory of relative deprivation suggests that anger and the desire for power play significant roles in fueling acts of political disobedience related to climate issues. In ecologically hostile regions, violent neo-Nazis have gained traction in modern history due to the failure of states to provide essential services. As a result of soil salinization and the transformation of grasslands into deserts, climate change has greatly impeded the capacity of VNSAs to enlist fighters and assert their authority in the realm of agriculture. Certain regions within African states, like Mali, Algeria, and Libya, are already practically unruly due to geography. In Mali and Nigeria, farmer-herder conflicts are becoming worse as a result of climate change, which is also causing desertification, destroying livelihoods, and driving migration. The emergence of VNSAs can be influenced by environmental pressures, even in more developed nations with better governance. There have been rebellions and tensions as a result of governments implementing necessary but invidious water conservation measures in response to a historic megadrought in the Western United States. More radical water politics and increased VNSA activity may be on the horizon as the megadrought in the American West intensifies and the federal government mandates water reductions for Western states. Many individuals in California express concern over the potential for armed militias to exploit forest fires as a means of recruiting new members. If states fail to meet the demands resulting from climate change, there will be a need for alternative solutions. In this context, the expansion of ungoverned territories can serve as fertile ground for the growth of these alternatives, particularly in areas where traditional crops cannot thrive (Kuyper, Linnér, and Schroeder, 2018).

Worries about climate change are prompting governments to impose stricter regulations on the extraction and burning of fossil fuels, which could spark violent protests. At least one riot related to public demand for fuel occurred in 41 countries between 2005 and 2018. More and more nations are reducing subsidies for fossil fuels and instituting carbon pricing in response to growing climate change concerns. Unfortunately, consequences of the shift to renewable energy sources and the escalating costs of fossil fuels have sparked early protests. Joining forces with groups like the Beyond Oil and Gas Alliance to end the extraction of fossil fuels has been a reluctance on the part of many governments and large fossil fuel producers. Governments will face pressure to limit production from oil fields and other "climate-forcing assets" like cattle farms if they are serious about meeting their climate commitments. Lithium, cobalt, copper, and rare earth minerals needed for renewable technology and wind turbines are among a new class of topical resources that are becoming more valuable and important as the world shifts to low-carbon energy. Wars have broken out in previously unanticipated places as VNSAs move to seize control of these resources. There will be a surge in the demand for border guards armed with weapons and people smugglers as millions of people are forced to flee their homes and countries due to climate change. As a result, the United States' southern border and the Western and Central Mediterranean migratory routes to Europe generate billions of dollars in revenue and over $100 million annually, respectively, from climate refugees who are unable to enter countries where they are legally not welcome. Smugglers are able to finance illicit armed activity because more people are giving them their savings. On the other side of the border, vigilante organizations like the 3% and the Minuteman Project could be ready to stop migrants. One possible response to climate change, for wealthy individuals with the proper passports, is to move to private communities that have their own power grids and desalination plants. Private security firms, like those in Los Angeles and Chicago, would have to install gates and armed guards to keep these communities safe. Governments spend more on post-disaster aid, which can lead to chaos and injustice following climate-related disasters. Examples of how disaster response can be inadequate include the 2010 floods in Pakistan and the coronavirus pandemic in Mexico, where "narcos" used medical aid to gain support.

A widening chasm now separates politicians' rhetoric about the "existential threat" and their actual responses to climate change. Activists may resort to violence in the face of intensifying climate impacts. They may attack fossil fuel infrastructure in an effort to reduce emissions and increase the cost of burning these fuels, or they may try to pressure politicians into action. Earth Liberation Front and Movement for the Emancipation of the Niger Delta are just two examples of groups that have used violence in the past to defend local communities and the environment. Climate protests, on the other hand, are often nonviolent, and many activists have resorted to minor acts of vandalism (Kuyper, Linnér, and Schroeder, 2018). With climate change expected to worsen over the next decade, some climate justice protesters may hold the belief that abducting high-ranking officials from fossil fuel companies would mitigate the problem. If a citizen feels wronged because their sick loved one did not make it through the most recent heatwave, they may resort to violent means to exact revenge on the politician they hold responsible. By 2040, the 2021 U.S. National Intelligence Estimate identifies significant risks to U.S. national security interests. These risks include the growing demands of developing countries for financing and technology assistance, as well as an increased need for aid and humanitarian relief (Bäckstrand et al., 2017).

7.7. THE ROLE OF NON-STATE ACTORS IN MONITORING AND REVIEW

The significance of non-state actors in the process of monitoring and review cannot be overstated. These actors play a crucial role in ensuring accountability and transparency in various domains. Their involvement brings a fresh perspective and diverse expertise to the table, enriching the overall monitoring and review process. By actively participating and providing valuable insights, non-state actors contribute to the effectiveness and legitimacy of these mechanisms. Their contributions are instrumental in fostering a more inclusive and comprehensive approach to monitoring and review (Nasiritousi, Hjerpe, and Linnér, 2016).

It has been demonstrated that non-state actors can assist governmental entities at the national and subnational levels with development program research, analysis, implementation, monitoring, and evaluation. Participation from non-state actors at various points in the development process can track and assess the achievement of two main goals are:

1. To start, they can make the system more open and accountable by getting more people involved and invested in the development process. To put it simply, non-state actors mediate between the people and their government. They raise people's consciousness by relaying official government messages to the general public, and they relay the public's reaction to those messages back to the government, so the latter can learn from both the good and the bad.
2. Second, by participating actively throughout the policy cycle, non-state actors can assess both the preparation and execution of policies. They can lend a hand and give each other a chance to assess how well they're doing in developing and executing development agendas thanks to their expertise in policy planning and execution.

The following chapters elucidate these responsibilities, discussing potential difficulties, monitoring, review procedures, and mutual benefits for state and non-state actors.

CHAPTER 8

THE SIGNIFICANCE OF PUBLIC AWARENESS IN CLIMATE SECURITY

Contents

INTRODUCTION

This is understanding the level of general awareness regarding conflicts, disasters, and risks, as well as the various factors that contribute to these events and the measures that can be implemented both individually and collectively to minimize exposure and vulnerability to hazards. Effective disaster and conflict risk reduction relies heavily on public awareness. The development of this initiative is achieved through various means, such as:

- The dissemination of information via media and educational platforms;
- The creation of information centers and networks; and
- The implementation of community engagement and advocacy efforts led by senior public officials and community leaders.

8.1. THE IMPORTANCE OF CLIMATE CHANGE AWARENESS

Raising public awareness is crucial for tackling global challenges such as climate change. One of the most important things the public can do to help climate change adaptation and mitigation efforts is to be aware of the problem. As a means of psychologically coping, people who are more exposed and cognizant of dangers may start to normalize risks. The polar ice caps are melting, which threatens animals and increases the risk of floods, among other obvious consequences of climate change. Because many people dismiss climate change as an international problem that has no bearing on their lives, or even deny its existence, its impact has been under-recognized (Figure 8.1).

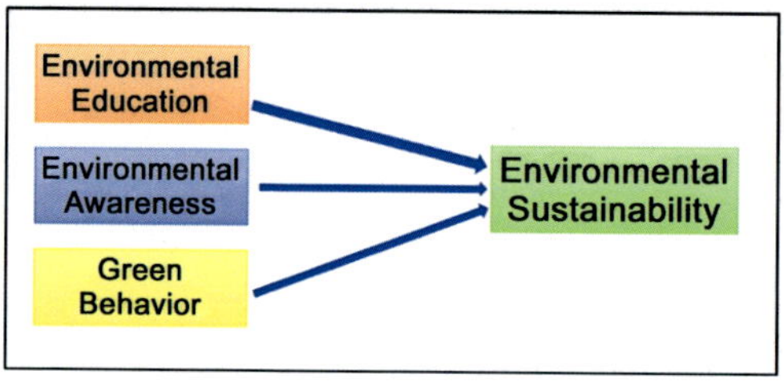

There has been a mismatch between public and scientific understanding of climate change. While government policies and planning are essential to this change, they are not enough on their own; this fosters denial and the spread of false information. Empowering the community to take action, raising awareness about the problem, and providing information about available solutions are all crucial. Get the people involved if you want your climate change plan to succeed (Climate-ADAPT, 2023).

Figure 8.1. *Environmental education for a sustainable future.*

Source: Yang K-F, Liu P-R, Zheng Y-H, Peng Y-J, Wang H-Y, Huang Y-F. Environmental Movement Regardless of Nationality: International Students' Attitude toward Environmental Protection and Environmental Awareness in Taiwan. Engineering Proceedings. *2024; 74(1):26. https://doi.org/10.3390/engproc2024074026.*

After languishing in the early 1980s, when the problem was first acknowledged, public concern, awareness, and understanding of climate change have surged due to mounting scientific evidence and increased media coverage (Knight, 2016). Citizens can play a larger role in decision-making when they are better informed about climate change, which can raise community-wide awareness and understanding. Changes in public attitude and action on climate change are possible outcomes of increased public knowledge, understanding, and engagement.

We can mitigate the worst effects of global warming by lowering the global carbon footprint. The environmental crisis may seem too far away for the ordinary person to be concerned about. On the other hand, people's daily habits can also affect the planet. In order to combat climate change and related issues like air pollution and waterway contamination, it is necessary to bring more attention to these environmental crises.

Regular people can help the environment and fight these problems by doing little things. The Climate Crisis Film Festival (CCFF) is one example of a multi-form approach; it was the first of its kind in the United Kingdom. Its primary goals include bolstering youth-led programs and grassroots solutions to the climate crisis and raising global awareness of the issue through innovative means (WHO, 2024; Figure 8.2).

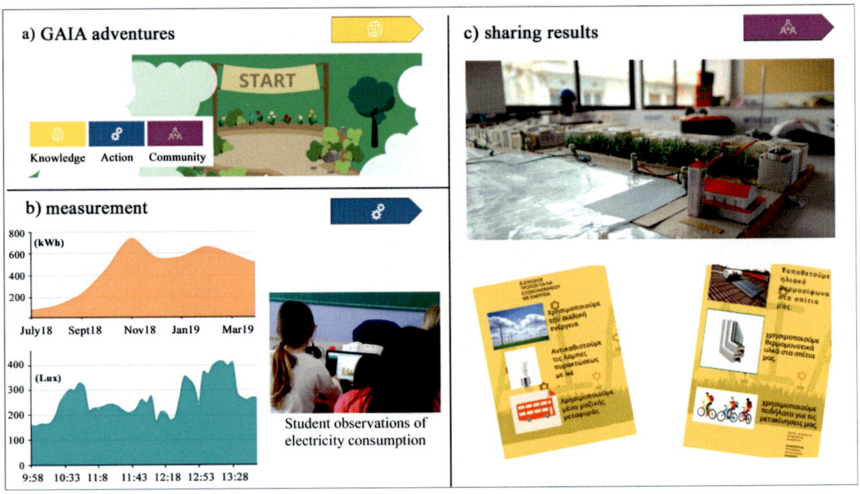

Figure 8.2. *Awareness programs.*

Source: Maurer M, Koulouris P, Bogner FX. Green Awareness in Action—How Energy Conservation Action Forces on Environmental Knowledge, Values, and Behaviour in Adolescents' School Life. Sustainability. *2020; 12(3):955. https://doi.org/10.3390/su12030955.*

Instead of focusing on adaptation, many massive climate change awareness campaigns promote energy efficiency, sustainability, and mitigation. For instance, the You Control Climate Change (2006) campaign by the European Commission aims to educate people about climate change, start proactive conversations, and encourage (small) behavioral changes that won't affect people's daily lives, all while making them feel empowered and responsible. The Commission has been attempting to mitigate the skepticism of the targeted group by highlighting the importance of individual actions in combating climate change. Advertising, a website, exhibitions, media relations, events, and national and European school programs have all been heavily invested in by the EU to achieve this goal (Knight, 2016).

8.2. THE ADVANTAGES OF INCREASING AWARENESS

Increased awareness fosters greater enthusiasm and support, encouraging individuals to mobilize themselves and take action. Awareness raising plays a crucial role in climate change adaptation as it helps manage the impacts of climate change, enhance adaptive capacity, and reduce vulnerabilities. The objective of any climate change campaign is to foster lasting changes in behavior that result in positive outcomes. These changes are brought about through the active involvement of the public. Increased awareness heightens one's level of alertness, ultimately prompting individuals to take action. After grasping the core of the issue, one can proceed to assess which alterations to their regular routine would promote environmental friendliness. This person demonstrates a clear understanding of the potential negative consequences of certain actions and takes proactive steps to minimize their carbon footprint. By fostering climate literacy among young people, it empowers them to transform their attitudes and behaviors while equipping them to navigate the challenges posed by global warming (Climate-ADAPT, 2023).

8.3. WHAT PREVENTS A PERSON FROM ACTING?

Public support and engagement in climate change initiatives are impacted by how people perceive the risks involved. Even with the various awareness approaches used, not many changes have occurred. The normalization of risk may explain why people with a greater understanding of climate change perceive fewer risks.

Gradual adaptation causes risky or hazardous practices or conditions to become acceptable over time. This has an effect on how individuals perceive and comprehend global warming. According to Knight (2016), there is a divide between the theoretical concept of climate change and its practical application because individuals who are aware of the issue do not incorporate its psychological, political, or moral implications into their daily lives or social actions (implicatory denial). Additionally, some individuals are aware of climate change but mistakenly believe that it is harmless or won't have a significant impact (interpretive denial).

Most people will feel the impacts of climate change in the far future. According to the construal level theory, individuals tend to conceptualize things more abstractly when they are psychologically distant from them, whether in terms of time, space, or social distance. Wildfires and other severe weather events, which are likely caused by climate change, often strike remote areas, far from the homes of the majority of the population. As a result, most people can just think about climate change as a general idea and avoid having to deal with the details. When it comes to abstract concepts, people tend to be less forceful.

Another obstacle is the cost. For financial reasons, many people cannot purchase environmentally friendly alternatives. Many people have been hesitant to make the transition from gas-powered vehicles to electric ones because of the high cost of such a major life change (WHO, 2024). Although there are many challenges, there are also many solutions to this problem. All it takes is a simple online post or two to raise awareness about climate change. To make an impression, minor adjustments are sufficient. Because everyone has their own personal obstacles, even baby steps can make a difference. One can be resourceful in their approach to raising awareness, which is a wonderful way to help. Posting a short film or piece of art online has the potential to reach millions of people, allowing for the gradual spread of information. Having access to that data can be a powerful catalyst for change.

8.4. AWARENESS RAISING CAMPAIGNS TO CHANGE STAKEHOLDER BEHAVIOR

An "awareness raising campaign" is a general term for a set of coordinated efforts to inform a specific demographic over a set amount of time. While the specific goals of awareness-raising campaigns may vary depending on the setting, they often center on informing the intended audience about the issues at hand and offering solutions or behavioral changes to alleviate or eliminate them. Although it is commonly believed that awareness raising is crucial in the early stages of adaptation, studies have shown that levels of awareness change over time due to environmental factors (Knight, 2016). In light of this, it is essential to raise awareness not just in the beginning but all the way through.

Climate change awareness campaigns can target communities impacted by a specific risk, as well as stakeholders, companies, and the general public. The end goal of these kinds of campaigns is to get people to change their behavior in a way that stays. People, organizations, and decision-makers are better informed when they are made more aware of a problem. Its stated goal is to have all applicable sub-regional and regional organizations learn about the effects of climate change and prepare for them. As an example, "The Netherlands Live with Water" raised awareness about the dangers of coastal and river flooding; however, such campaigns can also zero in on a particular impact that is deemed most pressing for a particular location.

By employing a range of communication strategies, the effectiveness of awareness campaigns can be greatly enhanced. These strategies include:

- Organizing public meetings and training sessions;
- Consulting with professionals;
- Distributing printed materials;
- Utilizing social and mass media platforms; and
- Tapping into informal networks.

Case studies like Vrijburcht, an Amsterdam climate-proof collective garden that was privately financed, show how community self-protection teams can work in tandem with awareness campaigns to reduce the likelihood of harm coming to individuals and businesses alike (for instance, in the event of a flood) (Figure 8.3).

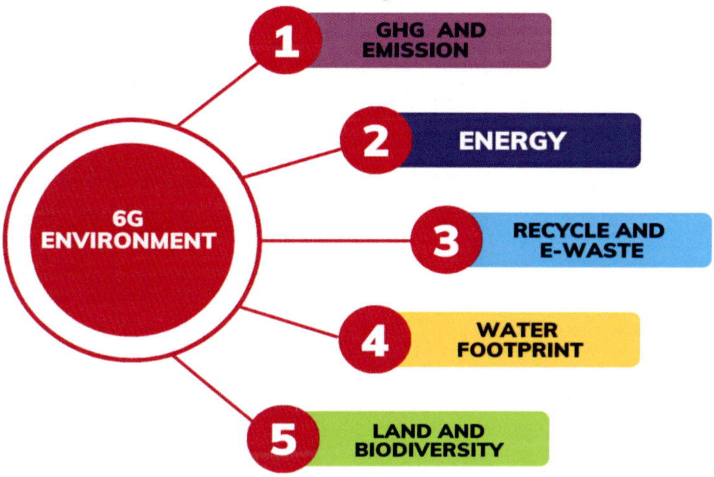

Figure 8.3. *Environmental awareness programsss.*

Source: Chauhan D, Mewada H, Gondalia V, Almalki FA, Patel S, Modi H, Kavaiya S, Trivedi Y, Mujlid HM. Balancing Technological Innovation and Environmental Sustainability: A Lifecycle Analysis of 6G Wireless Communication Technology. Sustainability*. 2024; 16(15):6533. https://doi.org/10.3390/su16156533.*

The message can be conveyed through a variety of media, such as television, the internet, social media, and newspapers, among others. Not only that but there are several resources out there that aim to raise consciousness among both decision-makers and the general public, including the Change Game and the European Union's communication campaign on climate action. Instead of solely emphasizing adaptation strategies, large-scale campaigns to raise climate change awareness typically incorporate a mix of mitigation, energy efficiency, and sustainability measures (WHO, 2024).

8.4.1. Engagement of Relevant Parties

Everyone from individuals to governments to corporations to industries is going to have to work together to adapt to the effects of climate change. Environmental non-governmental organizations (NGOs) and other relevant stakeholders can greatly improve the efficacy of awareness campaigns by contributing to their creation and implementation. They have a deeper understanding of "their" clientele and how to effectively reach out to them. Involving them also gives you more leverage and makes your campaign seem more credible.

8.4.2. Reaching Goals and Overcoming Obstacles

The effects of climate change, potential strategies for adaptation, the relative costs and benefits of various actions, and the general public's level of knowledge on the matter are not always well-communicated. A campaign to raise awareness can help resolve these problems. The following are crucial aspects of their success:

- When determining the target audience, it is crucial to assess the individuals who are most vulnerable and those who have the most to gain;
- Utilizing clear messages is crucial for capturing the attention of the target audience;
- Messages should present a compelling argument for personal motivation;
- It is important to communicate messages in the language that the audience is most comfortable with;
- When crafting messages, it is important to focus on the potential gains or losses that may arise from adaptation;
- It is important to provide precise instructions on how individuals can effectively address the specific risk at hand;
- It is crucial that messages maintain a sense of accountability and ownership at all levels;
- The communication strategy should be tailored to the specific audience, such as young people who can be reached through online platforms; and
- Finally, it is crucial to thoroughly evaluate the campaign model and communication modalities.

Nonetheless, the involvement of stakeholders and the effectiveness of the campaign may face obstacles due to the potential lack of awareness and understanding regarding climate change, leading to a perception that the issue is remote and detached from reality.

In addition, the success of the campaign could face obstacles due to potential underrepresentation in political agendas and the perception of helplessness resulting from the lack of action on climate change at the local, regional, and national levels.

8.4.3. Cost and Benefits

Gaining the attention of individuals is a challenging task, and the results can be quite unpredictable. Assessing the effectiveness of awareness campaigns can be quite a complex task, as there is often a lack of clear indicators to measure their impact. However, conducting surveys, both quantitative and qualitative, can yield valuable insights and information. Expenses related to a campaign involve more than just creating and distributing campaign materials. They also include the strategic planning and implementation of captivating activities. It is important to take into account the desired audience size and the chosen method of content distribution when considering these expenses (work cost per person) (Knight, 2016).

8.4.4. Commencement Date of the Project

Establishing a clear timeline for the launch of awareness-raising campaigns is crucial, given the complexity of coordinating multiple outreach events and involving various stakeholders. For successful outreach strategies and activities, it is crucial to allocate a sufficient timeframe of one to five years for implementation.

8.4.5. The Entirety of One's Existence

The duration of an awareness raising campaign can greatly differ depending on its scale. Their contributions are aimed at creating lasting impact, as they are crafted to drive meaningful societal shifts and foster the adoption of sustainable practices.

8.5. WAYS TO RAISE CLIMATE CHANGE AWARENESS AT A PERSONAL LEVEL

Global warming is among the most pressing problems our world is currently experiencing. There are many ways to keep people informed, even though some may not realize the gravity of the situation.

8.5.1. Post News on Social Media

It is important to disseminate information through various online platforms, including social media. Given the prevalence of online information consumption, utilizing this platform can effectively capture widespread attention to the issue at hand. It would be beneficial to explore online resources like videos, news articles, and recent research on climate change. Make sure to rely on reputable sources like the EPA, the NOAA, or NASA. Share the links on your social media platforms and encourage others to do the same. Consider searching for video clips that are approximately 30–45 seconds in duration, as individuals tend to be more inclined to watch shorter content rather than read an entire scientific article. Academic institutions and journals are reliable sources of information due to their rigorous research process. It is crucial to thoroughly review an article and verify its source before sharing it to prevent the dissemination of any inaccurate information.

8.5.2. Place Awareness Posters on Display

With some basic signage, you can reach the people in your community. There are many ready-made posters available online or you can create your own. Put some data, pictures, or infographics on your poster to support your points—for example, projected sea level rise or glacier melting. Place the posters around your neighborhood and in areas that see a lot of foot traffic. You could mention, for instance, that if human emissions continue, the hottest days of the year will feel 10–15° hotter by 2100. Another example would be to write that 20–30% of wildlife species may become extinct if temperatures continue to rise and include images of these animals. You can print off free images and graphics from the websites of many climate change organizations to use for your posters. As they might have different policies, make sure you get permission before hanging posters at places of business or education (Climate-ADAPT, 2023).

8.5.3. Make Artwork that Addresses Climate Change

Using your work, showcase the consequences and visuals of climate change. Since art inspires so many people, using it to express your concerns about climate change might be a good idea. You can use any media that you believe best expresses your message, including paintings, collages, photos, sculptures, and more. To ensure that the person viewing your work understands exactly what you're trying to convey, include an artist's statement explaining how climate change inspired it. For instance, you can alter a picture so that a well-known landmark appears to be underwater as a result of rising sea levels. For another example, you could recycle plastic or trash to create a collage that illustrates the amount of waste an individual produces on a daily basis.

8.5.4. Make Talks Regarding Climate Change

Put your knowledge into a simple visual slideshow. You can discuss climate change in general terms or delve deeper into specific effects. Conduct extensive research to identify trustworthy sources, then incorporate the data into your presentation. To maintain engagement and interest among those you are presenting to, incorporate a few pictures and information on each of your slides.

When you are presenting, remember who your audience is. Use relatable, basic topics when presenting to younger children, such as how picking up litter maintains the playground tidy and removes hazardous waste from the environment.

8.5.5. Discuss Finding Worldwide Solutions to the Issue

To show someone that change is possible, concentrate on finding solutions to problems. There are a ton of people working to improve our situation, even though climate change can be terrifying and seem inevitable. Informing the person about the initiatives being taken by nations and communities worldwide to assist will give them hope. This may give them the realization that they are not helpless and that they can take action to address the issues. For instance, you could discuss how nations are switching to renewable energy sources like solar and wind power. To cut down on carbon emissions, you can also bring up global campaigns like the Turn It Off campaign, which encourages people to switch off their cars' engines rather than leave them running.

8.5.6. Talk About Your Individual Climate Change Mitigation Efforts

Inform individuals about proactive measures they can take at home. Everyone can slow down climate change and lessen their carbon footprint by doing a plethora of different things. Discuss with them some easy changes they can make, such as using energy-efficient lights, recycling, composting, and carpooling. Inform them that single-use items generate a significant amount of waste and provide guidance on finding durable, reusable substitutes. A few more things you can do at home are purchase energy-approved appliances, insulate your walls, and practice water conservation.

8.5.7. Organize a Volunteer Initiative

Organize a community event to help spread the word locally. See if there is anything you can do to help spread awareness of climate change by having a conversation with the folks in your school or neighborhood. You may start a community garden, plan a park clean-up day, or have a bake sale with the proceeds going to a climate organization. Inform people of the project's environmental benefits or the charities you plan to support with the money you earn.

8.5.8. Joining a Community Climate Committee as a Volunteer

Spend some time supporting the environment in your community. Look for organizations in your city or neighborhood that are presently making improvements to the area around you. They might work to rehabilitate your neighborhood's green areas, create wildlife habitats, and introduce climate-adaptable plants. Inquire about the best ways to assist so that you can lessen the environmental effects in your community and form relationships with like-minded individuals. If you are still enrolled in school, find out if there are any extracurricular or environmental clubs that you can join.

8.5.9. Make Contact with the Local Legislators

Share the importance of climate change with your representatives. Policies that reduce emissions and enhance the environment can be passed with the assistance of your local government. Choose candidates who support addressing climate change and have green policies. Inform your representatives in writing of your concerns and request action on climate change while they are still in office. To learn more about their policies and how to contact them, look up the representatives in your state and city. Politicians are generally happy to hear your concerns so they can better represent you, but we understand that calling or writing can be a little unsettling.

8.5.10. Cite Empirical Studies and Scientific Findings

Present your findings and facts to skeptics to win them over. Check if the most recent research you can locate are from reliable sources by reading them in a scientific journal or visiting an official website like the EPA that addresses climate change.

Discussing facts supported by science can be more persuasive when discussing climate change with others. People will be more receptive to your words if they know you are a trustworthy source, so be sure to share where you got your information. For instance, you may remark, "Did you know that Earth's average temperature could rise by an additional 8.6°F in the next 100 years?" Skeptics sometimes don't trust evidence or scientists. Speaking with those who share your belief about climate change may be a more worthwhile endeavor than engaging in conflict with denialists.

8.5.11. Take up the Role of Citizen Scientist

Keep an eye on local climate change to aid scientists. A lot of researchers ask people to record information about when plants bloom, the level of pollution in the water, or how weather changes affect wildlife. To find out how you can become involved in your community, get in touch with the local climate activists and organizations. Additionally, you can search for nationwide initiatives that pique your interest and find out if any local volunteers are needed.

8.5.12. Enroll in a Climate Action Petition

Participate in online fundraising for a cause. Online, there are a ton of petitions urging businesses and governments to take climate change more seriously. Look for petitions about climate change on websites and read through the proposed solutions. You should sign any causes you support and forward them to others so they can sign them as well. After signing a petition, you can share it on social media to let your friends and family know what you're supporting.

8.5.13. Give Money to a Charity

Donate money to those who can assist with broader issues. Numerous nonprofit organizations strive to alleviate environmental problems and raise public awareness of climate change. Donate what you can if you have extra money so they can continue to advance the battle against climate change.

8.5.14. Participate in a Demonstration or Rally

Speak with people in your community face-to-face about your concerns. Climate activists have organized a plethora of marches and rallies, demonstrating the broad public support for climate change. To find out when protests are scheduled in your area, follow the local community organizers. Large-scale protests always pique people's interest, so you could get more people interested in your cause and educated about climate change. Meet others in your community who share your concern for the environment by attending rallies.

8.5.15. Give Instances from Real Life

If someone is blind to the big picture, discuss the obvious effects. It's possible that some people are unaware of the consequences of climate change and how serious of an issue it is. Talk to them about the ways that climate change has affected your community so they can better understand the issues. You can also discuss how a person's interests in nature or travel have been impacted by climate change. For instance, you may discuss how the shifting climate in their native habitat has caused new insect pests to relocate to your area. If the person has a passion for flora and fauna, mention that 20%–30% of species face extinction due to climate change.

CHAPTER 9

CLIMATE CHANGE AND ETHICS

Contents

INTRODUCTION

Climate change is a pressing concern that demands our attention. Numerous scientists have warned about its potential to bring about detrimental effects on all living beings, extending beyond just humans. According to Broome (2008), the impact of human activities on Earth is causing severe harm to various life forms, future generations, and even our fellow humans.

Given the circumstances, it can be argued that the main purpose of ethics is to establish a structure for evaluating appropriate responses to encountered dangers and, to some extent, contribute to them. Every time someone makes or evaluates a proposal for a course of action, they are engaging in practical ethics. On the other hand, experts in academia tend to place greater emphasis on elucidating the structure of ethical considerations applied to a given issue and the underlying assumptions being employed.

Climate change poses a significant threat to human well-being, prompting us to delve into the ethical dimensions of how we should address this issue. We must carefully consider the appropriate and inappropriate responses to this pressing challenge. Understanding the allocation of responsibilities and its impact on the rights of individuals, including the right to be protected from the repercussions of climate change, is of utmost importance (Broome, 2008).

9.1. RESPONSIBILITIES ON CLIMATE CHANGE

When considering the collective actions of humanity, it becomes apparent that there is potential for the pursuit of specific goals aimed at reducing or capturing carbon emissions. In certain cases, it may also be necessary to implement measures to support individuals who must adapt to the consequences of climate change. Nevertheless, it is important to note that humankind is not a singular entity. One of the key topics in climate ethics revolves around the question of who bears the responsibility for taking action. There are typically two main aspects that are commonly discussed: mitigation and adaptation.

Both the decisions that contribute to climate change and the resulting effects can be subject to ethical evaluation. Certain philosophers highlight the differentiation between the two types of responsibility. Additionally, some scholars place significant emphasis on the concept of mitigation or adaptation. However, there are also compelling arguments in favor of an integrated theory.

The question of responsibility and who should bear the burdens has sparked intense debate. In terms of the various entities that can bear this responsibility, such as individuals, corporations, and states, the current focus has primarily been on nation-states. There has been an international consensus on the principle of common but differentiated responsibilities among them.

According to Grubb's observation in 1995, the primary concern has always been the fair distribution of emissions among states. While this assumption has faced criticism, it is important to consider the significance of states in contemporary global politics. They continue to hold significant decision-making powers. The main focus has been on the criteria used to differentiate responsibilities.Given the nature of responsibilities and people's tendency to avoid them, it is no surprise that there is a significant amount of debate on this topic. The issues at hand have been given special attention in relation to discussions surrounding global climate change agreements.

9.1.1. Grandfathering Principle

The Kyoto Protocol implemented the 'Grandfathering Principle,' which mandated that developed countries, also known as Annex-I countries, had to decrease their emissions by an average of 5% in relation to 1990 levels. Therefore, countries that were already significant polluters in 1990 were allowed to continue emitting higher levels of greenhouse gases than countries with lower emissions. During post-Kyoto negotiations, there is an ongoing discussion about including developing countries in the emissions reduction program. The wealthier nations are strongly advocating for the application of this principle. When examining the principle's rationale, philosophers have typically viewed it as a practical compromise rather than a moral justification. Even if it is the only politically feasible option, it is still preferable to forego an agreement altogether if it means getting major polluters to accept emissions reductions.

Recent efforts to explore the moral implications of this issue have not been able to fully address the primary ethical concerns. There are concerns regarding the grandfathering of emissions rights. One concern is that it could perpetuate existing inequalities by denying those who are worse off the same opportunities and quality of life enjoyed by affluent citizens of countries with higher levels of pollution. Another concern is that it could hinder less economically developed countries from addressing energy poverty, thus trapping them in a state of underdevelopment. There is a compelling ethical argument for the principle that individuals who have contributed to the excessive burden on the atmosphere should bear moral responsibility for addressing the resulting consequences. This principle resonates with many people on an intuitive level. However, there are certain factors related to fairness and feasibility that argue against the idea that moral responsibility should align with causal responsibility. One must consider the distinction between applying it to actions that are yet to occur or are currently happening, as opposed to applying it in hindsight. One could argue that applying the principle retrospectively is generally considered illegitimate (Broome, 2008).

9.1.2. Polluter Pays Principle

Specifically, the Polluter Pays Principle (PPP) is often used as a practical tool to address the externalized environmental costs of economic activities. It is not typically employed as a principle of historical accountability, although it is sometimes mistakenly associated with it. The principle aims to internalize these costs and hold those responsible for causing pollution accountable (Figure 9.1).

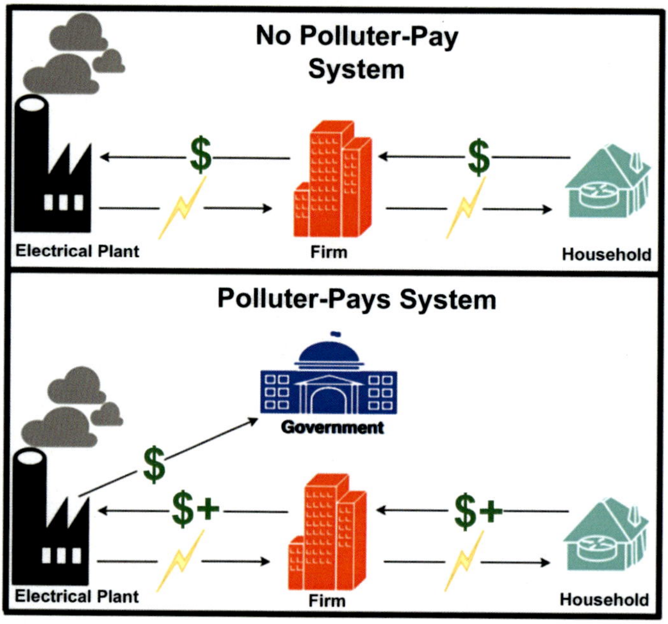

Figure 9.1. *Flow chart illustrates the impact of PPP cost on consumers/producers.*

Source: Energy Education, 2024.

It is important to grasp the concept that this principle is focused on the future, enabling logical decision-making regarding the acceptability of costs before they are incurred. Imposing obligations and costs in hindsight, which raises ethical and legal concerns, becomes even more problematic when considering that previous generations may have been unaware of the harmful consequences of their actions. In addition, it is important to consider the practicality of applying principles of liability to individuals who are no longer alive. There is an argument that suggests amending the causal responsibility principle to include the acceptance of liabilities when inheriting assets from our grandfathers.

According to Singer (2006), adopting this stance means not just making amendments, but completely abandoning the causal responsibility principle. It is a clear and distinct position to take.

9.1.3. Beneficiary Pays Principle

Another principle that is often discussed in the climate justice literature is the 'Beneficiary Pays Principle.' According to this principle, individuals who have benefited from past activities that have caused harm to others also have a responsibility to help alleviate those harms. In the Kyoto discussions, there is an approach known as 'the Brazilian Proposal' that suggests considering countries' historical emissions. This is important because countries that have been heavily polluted have actually benefited from it in terms of increased wealth and more developed infrastructure. These countries have a responsibility to repay their ecological debt (Broome, 2008).

An objection that has been raised is the argument that it is unjust to blame present members of wealthier nations for the excessive use of the atmosphere by their ancestors, especially when they had little control over the energy decisions made in the past.

On the other hand, some argue that residents of developed countries currently have numerous advantages compared to those in developing countries due to the emissions produced by their nations in the past. It is important to acknowledge that developed countries have benefited from past emissions and should not be exempt from taking responsibility for the negative consequences that accompany their high standard of living. Another point to consider is that certain advantages resulting from previous emissions extend beyond the borders of the countries responsible for them. Singer (2006), for instance, presents the argument that previous emissions have played a crucial role in enabling the creation of public goods, such as advancements in medicine and technology, which have significantly contributed to the improvement of living standards in developing nations. While less developed countries may have received some benefits, they have mostly been burdened with a significant amount of debt. This debt has been incurred in their pursuit of purchasing the products of industrialization. Another possibility to consider is that past emissions may not have provided significant benefits to the present generation. In this case, Implementing thresholds of ability to pay seems like a reasonable solution.

9.1.4. Seek Responsibility and take Responsibility for Your Actions

Another principle often discussed in the climate justice literature is the idea that those who have the means should take on the responsibility. There are practical benefits to this approach compared to the previous two, as it avoids the need for investigating causality or dealing with other challenging historical factors. Allocating responsibilities based solely on the ability to pay is generally not recommended. This approach can lead to injustice or create perverse incentives, especially when the ability to pay is a result of efficiency or ecological frugality. However, it can be argued that this approach is suitable in cases where the other principles are ineffective or inapplicable.

When discussing the allocation of responsibilities, it is important to consider how they are defined and the boundaries they create. Additionally, it is crucial to address the corresponding rights that individuals possess. As an example, consider the concept of determining a responsibility to limit CO_2 emissions. This essentially involves granting a license to emit up to a certain amount of CO_2. In practice, negotiators are primarily interested in knowing this specific emission limit (Broome, 2008).

9.2. DISCUSSIONS SURROUNDING RIGHTS

Disputes about climate ethics have made heavy use of the language of rights. Specifically, the question of whether and how climate change can be considered a human rights issue is a matter of contention. There are two contrasting perspectives on this matter are:

1. One argues that it is crucial for humans to benefit from the planet's carbon absorption capacity, as it is a valuable resource; and
2. Another viewpoint suggests that protecting humans from harm is paramount, as it can jeopardize essential human needs.

An equal per capita entitlement as a suitable metric for international agreements has arisen alongside the concept that humans have a right to a specific level of greenhouse gas emissions. The right to equal or reduced emissions has been advanced by some.

Since its original proposal by Henry Shue—albeit in relation to ozone emissions—the concept of a human right to "subsistence emissions" has played a significant role in discussions regarding the moral implications of climate change. But other philosophers have weighed in with doubts. People have the right to whatever is required to meet their basic needs as humans, but they do not have the right to choose how those needs are met (Singer, 2006). Additionally, carbon emissions should not be considered a human right because they are not necessary for a decent human life. Sustenance needs can be satisfied (and have been for most of human history) without economic development driven by fossil fuels. Instead of solving the problem of excessive emissions, a right to minimum emissions runs the risk of making it worse. Instead of viewing the wealthy's responsibility to the poor as "more emissions," they should view it as a fair distribution of the advantages they have enjoyed from their own excessive consumption of air, water, and other natural resources, collectively referred to as ecological space.

One aspect that is often highlighted in climate ethics discussions is the impact of climate change on the interests that human rights aim to protect. This helps to emphasize the importance of considering human rights in the context of climate change. Astonishingly, anthropogenic climate change violates human rights. People have the right not to be exposed to dangerous climate change. Not all philosophers believe that rights necessarily protect people's interests; and even if they can, it is not enough for there to be a right if there is an interest; this is why these arguments aren't always as clear as they could be regarding the relationship between interests and rights.

To have any weight, the assertion of a right must imply the existence of a responsibility to achieve the result that the right itself specifies. A strong allegation that someone is doing something direct and harmful is certainly implied when one speaks of a right being violated. However, in practice, accountability is more diffuse and multi-faceted.

The human rights discourse continues to gain momentum, even though this broader issue impacts rights unrelated to climate change as well. With procedural rights firmly established, there has been progress in recent years toward the international recognition of environmental human rights; however, there is also increasing recognition of a fundamental substantive right to an adequate environment. As a natural extension of this right, Vanderheiden has proposed a claim to climate stability; Hiskes maintains that environmental human rights can form the bedrock of environmental justice across generations. While there is a valid rationale for utilizing human rights discourse and tools, there remain inquiries and apprehensions regarding the application of human rights to environmental causes or the pursuit of environmental justice. When we delve into the long-term implications of climate change, these challenges become even more evident (Broome, 2008).

9.3. EXPLORING CLIMATE CHANGE AND ETHICS IN THE FUTURE

While it is true that we currently recognize the impact of climate change in the present, it is important to acknowledge that our attention is still heavily directed towards the future. The concept of justice between generations and our responsibilities to future generations are significant considerations that cannot be ignored.

The issue of climate change revolves around the notion that it is problematic to let the environment degrade for the sake of future generations of humans. However, upon closer examination, we are faced with certain perplexities.

9.3.1. The Concept of Non-Identity Problem

It is intriguing to consider that the wrongness of an act stems from its potential harm to future individuals or the negative impact it may have on their well-being. However, it is important to note that a person's identity is shaped by the time of their conception. Additionally, the choices we make in this generation will lead to the birth of different individuals in future generations. Consequently, it is not accurate to claim that we have caused harm or made future individuals worse off through climate change damage, as they would not have existed without the policies in question.

Various philosophers have different perspectives on how to address this complex issue. Some question whether it truly encompasses all the aspects we should be focused on, especially when considering a more holistic approach to the problem. However, it is important to acknowledge that there will always be a certain level of uncertainty when considering the ethical responsibilities we have towards future generations.

9.3.2. Assigning Rights to Future Generations and Taking a Non-Individualistic Approach.

We encounter a challenge when it comes to attributing rights to groups, even in the present. Furthermore, it becomes problematic to discuss the rights of beings who have not yet come into existence. Furthermore, these challenges cannot be overcome by merely discussing our responsibilities in relation to them, as similar inquiries arise. However, it is a common understanding that even though the future is filled with uncertainty, we must still make present decisions.

The concept of a stewardship approach to environmental ethics emphasizes the importance of treating the world as a trust, ensuring that we leave it in a condition that is not worse off for future generations.

The concept of sustainable development, which has been influential since the Brundtland Report 25 years ago, emphasizes the importance of development that meets the needs of the present while also considering the ability of future generations to meet their own needs. There has been minimal opposition to this concept, despite the raised concerns about why we consider the present as the benchmark for future generations when previous generations did not always inherit as much as we have. However, it is widely known for being open to various interpretations.

The question of how much of nature's 'capital' can be replaced by manufactured goods or other forms of technological capacity has long been a subject of debate in environmental values. The concept of 'strong' sustainability advocates for significantly less substitution compared to 'weak' sustainability.

The degree of constraint implied for economic 'business-as-usual' can vary significantly depending on the interpretation adopted (Singer, 2006).

As the constraints become less stringent, there is greater room to believe that humanity can overcome environmental issues by leveraging innovation and technology. This conception aligns with economists' assumptions regarding the potential for perpetual economic growth in the future. When discussing climate change, it is important to recognize that economic factors play a significant role in the debate. However, it is crucial to understand that these terms are not ethically neutral.

Frequently, this issue is discussed in the context of weighing the potential advantages for the economy against the potential drawbacks for the environment. However, it is important to consider the potential economic benefits of addressing climate change. This aligns with a long-standing tradition in the field of environmental economics, which aims to assign a value or price to environmental "services" that are currently unpriced or undervalued.

The influential report prepared by the British economist Nicholas Stern is a remarkable example. Concerning the future aspect, Stern devoted considerable thought to the ethical considerations surrounding the selection of the discount rate. This rate determines how the value of assets, welfare, costs, and economic damages, including those caused by climate change, are expected to decrease over time, from the perspective of the present. Stern's approach to determining the current 'correct' price for carbon emissions differed significantly from William Nordhaus's influential position. Stern used an overall social discount rate of only 1.4%, whereas Nordhaus employed a rate of 5.5%.59 Stern's analysis led him to advocate for the implementation of immediate emissions reductions, which would be in line with a carbon tax of USD300 per ton. In contrast, Nordhaus advocates for a carbon tax starting at approximately USD30 per ton and gradually increasing to USD85 per ton by 2050. The disparity in results can be attributed to the varying assumptions utilized in the computation of the social discount rate. The high rate suggests a significant lack of concern for future generations (Dietz, Hepburn, and Stern, 2007).

The economics of climate change cannot be divorced from the ethical considerations at play. It inherently involves making ethical judgments about our responsibilities towards future generations. Stern has faced criticism from various perspectives regarding the assumptions he permits. Both mainstream economists and more radical thinkers have recognized a politically-charged vision of a regulated capitalism, where it is not completely unrestricted but also not significantly questioned, continuing indefinitely in the future (Broome, 2008). Some liberal political theorists propose that we should adopt John Rawls's concept of a 'just savings principle' as a way to address our future obligations. However, it is worth considering what we can actually save for the future when we are already depleting the planet's resources and exceeding its environmental limits.

When we shift our perspective from expecting gradual environmental deterioration and incremental hardships to anticipating sudden and catastrophic changes, the nature and shape of the ethical problem before us undergo a significant transformation. Considering the potential consequences of runaway climate change triggered by reaching critical "tipping points," policymakers must address the crucial question of when and if a precautionary approach should be implemented.

The Precautionary Principle lacks a universally agreed-upon definition or a specific set of criteria to direct its application. However, efforts have been made to establish a fundamental concept for it through discussions on the precautionary approach to addressing climate change. It is important to consider the potential for harm from an activity, especially when there is uncertainty about the magnitude of impacts or causality. In such cases, it is advisable to take anticipatory action to prevent any harm.

According to Catriona McKinnon, the Precautionary Principle can be interpreted in different ways. While a strong interpretation may be too strict for most policy areas, McKinnon suggests that it could still be justified when it comes to climate change. If the potential consequences of not taking precautionary measures are so severe that they would make it impossible to achieve justice, then we cannot afford to ignore the principle (Singer, 2006).

9.3.3. Precautions to be Taken

There is a connection between climate change and the potential threat it poses to the conditions necessary for justice to thrive. The implications of environmental insecurities for ethical life have sparked a wide range of views. However, one overarching insight that has emerged is that the insecurities currently affecting people are further intensified by environmental changes, regardless of what the future may hold.

In various regions across the globe, the conditions necessary for justice to thrive have been eroded due to ecological marginalization, which is sanctioned and even promoted by the institutions that uphold globalization in the present day. One could argue that to ensure justice in the future, it is essential to wholeheartedly prioritize justice in the present. It is important to consider the long-term consequences and the needs of the most vulnerable in society. To achieve greater equality and address environmental challenges, those who are better off must be willing to make significant commitments (Dietz, Hepburn, and Stern, 2007).

From this perspective, climate ethics is considered an integral component of global justice, which is broadly understood. It is important to note that individuals who are already facing socio-economic disadvantage on a global scale are particularly vulnerable to the exacerbating impacts of climate change.

One could argue that the international institutional order provides a structural foundation for this significant disparity. These institutions are interconnected with our lives, as members of privileged societies, as emphasized by prominent justice theorists like Thomas Pogge. However, Pogge has faced criticism regarding the issue of individual accountability.

9.4. INDIVIDUAL OBLIGATION AND COLLECTIVE ACTION ON CLIMATE CHANGE ISSUE

The concept of individual obligation and collective action is a fundamental aspect that needs to be thoroughly understood and appreciated. It is crucial to recognize the significance of both personal responsibility and the power of working together as a group. By acknowledging our individual obligations and actively participating in collective actions, we can make a substantial impact on society and bring about positive change.

Figure 9.2. *Personal climate change action.*

Source: https://diff.wikimedia.org/2023/05/11/spreading-environmental-insights-in-the-igbo-community-africa-environment-wikifocus/africa-environment-wiki-focus-campaign-at-unizik/.

A significant portion of the literature on climate ethics focuses on the actions that should be taken by governments and policymakers. However, it is important to consider our individual responsibilities. According to Walter Sinnott Armstrong, the issue is mainly political and should be addressed at that level. In contrast, Jamieson presents a different perspective. According to him, while the nation state does play a role in addressing climate change due to its causal effectiveness, it is not the main entity responsible for ethical obligations He strongly emphasizes the importance of instilling green virtues at the individual level (Figure 9.2).

According to Singer (2006), it is crucial to examine the connection between individual and collective responsibilities. It is important to consider that determining the appropriate course of action can be complex. This is because it can be influenced by the actions of others, and individuals may have varying perspectives on what should be done, as well as differing levels of dedication to actually carrying it out. In addition, tackling a challenge such as climate change necessitates the coming together of various stakeholders. However, the question arises: how can we accomplish this when different individuals and groups are driven by divergent interests? In addition, it is important to consider how an individual's ethical obligations may be influenced by the actions of others who fail to fulfill their own responsibilities (Broome, 2008).

The collective action problem discussed in climate debates is of great importance. When individuals neglect their responsibilities, is it fair to expect others to shoulder the burden they have left behind? It could be argued that expecting individuals with strong moral values to shoulder the burden may not only be an unfair expectation, but it could also inadvertently encourage others to contribute even less. Therefore, it might be counterproductive, or even incorrect, for someone to engage in actions that they believe are universally expected. According to Singer (2006), outlining individual climate duties and creating universally applicable rules continues to pose a significant philosophical challenge (Figure 9.3).

Figure 9.3. *Collective climate change action.*

Source: https://en.wikipedia.org/wiki/File:Ende_Gel%C3%A4nde_November_2017_-_Front_banner_of_demonstration.jpg.

Emphasizing the importance of virtues, in addition to ethical principles like duty, rights, or utility, and considering ideas such as carbon allowances or allowances for various ecological services, is crucial when considering the actions individuals should take. One can envision a life that is abundant in experiences and uncomplicated in resources, where the emphasis is on existence rather than possessions. This way of living allows for a lighter ecological footprint and promotes compassion towards both humans and other living beings. If everyone were to engage in this behavior, it is possible that the issue of climate change could be completely mitigated.

There are, however, several obstacles that hinder the realization of this green vision. One aspect to consider is that not everyone will find it appealing. Another point to consider is that individuals who find it appealing often find themselves trapped in consumer-driven lifestyles. It is important to consider not only the desires and actions of individuals but also the institutional structures that uphold the incentives and motivations to sustain the current development path. Furthermore, it is important to acknowledge that people around the world are currently facing a wide range of circumstances. While some individuals may need to moderate their impact on the planet, others are simply striving to secure a basic existence. An impactful green vision would need to encompass a thorough examination of the structures of global inequality (Dietz, Hepburn, and Stern, 2007).

Climate change poses challenges of collective action, not just for individuals, but also for nation-states as a whole. In addition, there exist various types of collectivities, particularly those related to transnational enterprise and finance, which currently operate independently from state control. Given the profound influence that access to financial resources, employment within the financial system, or exclusion from global financial networks can have on an individual's life, modern ethics must address the interconnections among these various groups of people.

CHAPTER 10

FUTURE SCENARIOS FOR CLIMATE CHANGE

Contents

INTRODUCTION

Unless we significantly reduce the billions of tons of emissions we release each year, atmospheric concentrations of greenhouse gases will keep rising. The rise in concentrations will have several effects: the average temperature of Earth will increase; precipitation patterns and amounts will be influenced; ice, snow cover, and permafrost will decrease; sea levels will rise; the acidity of the oceans will increase; extreme events will become more frequent, intense, and/or longer-lasting; ecosystem characteristics will shift; and there will be an increase in threats to human health.

These changes will have far-reaching consequences on our:

- Food supply;
- Water resources;
- Infrastructure;
- Ecosystems; and
- Ultimately, our health.

The National Renewable Energy Laboratory (NRC) has a mission to stabilize the climate by decreasing the levels of greenhouse gases in the atmosphere.

Various factors, including volcanic eruptions, changes in solar radiation, and natural climate processes like shifts in ocean circulation patterns, can have an impact on the climate. In order to enhance our comprehension of these matters and make projections about forthcoming climate shifts, scientists employ computer models of the climate system (Kattsov et al., 2005).

10.1. GREENHOUSE GAS EMISSIONS IMPACT FUTURE CLIMATE FOR GENERATIONS

Over long periods of time, the atmosphere holds onto various greenhouse gases. As a result, the levels of greenhouse gases in the atmosphere would continue to increase and remain elevated for centuries, even if emissions were to stop increasing. It is important to note that surface air temperatures will continue to rise, even if we are able to stabilize concentrations and maintain a constant atmospheric composition.

Achieving this would require a substantial reduction in the emissions of greenhouse gases that are currently being used. The reason behind this is that the oceans, acting as a heat sink, require several decades to fully respond to the elevated levels of greenhouse gases.

The long-lasting impacts of climate change are inevitable, extending far into the future. This is due to the ocean's response to rising temperatures and increased greenhouse gas levels (Cubasch et al., 2001). Scientists utilize a range of scenarios that consider various assumptions about future economic, social, technical, and environmental conditions. This enables them to make predictions about long-term human-induced climate-altering factors, such as emissions.

10.2. FUTURE TEMPERATURE TRENDS

Over the course of the past few decades, there has been a noticeable increase in global temperatures. There are expected to be further temperature fluctuations in the coming years. Climate models project significant changes in temperature (Kattsov et al., 2005).

10.2.1. Notable Worldwide Predictions

According to projections, it is anticipated that average global temperatures will increase by 0.5°F–8.6°F by 2100. However, it is important to note that the minimum increase of 2.7°F is expected for all scenarios except the one that imposes the strictest limit on greenhouse gas emissions. It is projected that the global temperature will increase at a rate twice as fast in the upcoming century compared to the previous century unless significant measures are taken to mitigate the effects.

Based on predictions, it is anticipated that air temperatures in land areas will increase at a faster pace compared to those in oceans. There will be variations in temperature increases across different regions of the world.

10.2.2. Significant U.S. Forecasts

In the year 2100, there is an expectation that the average temperature in the United States will increase by around 3°F–12°F. This variation will depend on the emissions scenario and climate model being considered.

As global average temperatures continue to climb, the frequency and intensity of extreme heat events, commonly referred to as heat waves, are on the rise.

According to projections, the United States is expected to see a rise in the number of days with temperatures exceeding 90°F, especially as we approach the end of the century. In this particular instance, it is important to note the following: Based on climate models, it is projected that the United States will have summers that rank among the top 5% in terms of heat from 1950–1979 for at least 70% of the time between 2035 and 2064, assuming a continued increase in global greenhouse gas emissions (Cubasch et al., 2001).

10.3. PREDICTIONS FOR FUTURE PRECIPITATION AND STORM OCCURRENCES

There may also be changes in the frequency and distribution of storm events and precipitation types, such as snowfall and rain. Certain shifts are less uncertain than temperature-related shifts. Projected changes in precipitation and storms indicate variations based on seasons and regions. Precipitation patterns can vary across different regions, resulting in varying levels of rainfall. Some areas may experience a decrease in precipitation, while others may see an increase. Additionally, certain areas might not experience significant changes in their precipitation levels. In the near future, it is expected that there will be an increase in the intensity of downpours worldwide. Additionally, storm paths are projected to shift towards the poles.

10.3.1. Some Predictions on a Global Scale

According to Kattsov et al. (2005), by the end of the century, there is an expectation that the average annual precipitation worldwide will increase, although the extent and intensity of this change will vary significantly across different regions. Based on historical data, it is highly probable that precipitation events will increase in intensity. It is worth noting that the increase in overall rainfall is expected to be particularly pronounced in tropical and high-latitude regions. Anticipate an intensification of winds associated with the tropical storm. It is worth noting that tropical storms are anticipated to generate a greater amount of precipitation. There will be variations in annual precipitation across different regions.

10.3.2. Important Forecasts for the United States

In the northern regions, there is an anticipated rise in precipitation, especially in the winter and spring seasons. There is a projected decrease in precipitation for the Southwest and other southern regions. It is anticipated that there will be an increase in the frequency of heavy rainfall events, even in areas where the total precipitation is projected to decrease.

By the year 2100, there is a projected increase in the frequency of heavy rains. Currently occurring approximately once every 20 years, these intense rainfall events are expected to happen twice as often, or even up to five times as often, depending on the specific location. In the majority of areas, it is expected that there will be a greater amount of rainfall rather than snowfall, except in the most northern regions.

With the warming of the ocean, it is anticipated that the intensity of hurricanes in the Atlantic will increase. Climate models indicate that there will be an increase in the frequency and intensity of hurricanes, especially those categorized as Category 4 and 5. Additionally, these storms are expected to generate more substantial amounts of rainfall.

Winter storms are expected to continue their northward trajectory. It is anticipated that there will be a rise in the frequency and severity of winter storms (Cubasch et al., 2001).

10.4. ANTICIPATED ICE, SNOWPACK, AND PERMAFROST IN THE FUTURE

The melting of sea ice in the Arctic is already underway. Since approximately 1970, there has been a consistent decrease in the snow cover in the Northern Hemisphere. The permafrost in Alaska and a significant portion of the Arctic has experienced warming over the past century. The persistence of permafrost melting, glacier shrinkage, reduced snow cover, and declining sea ice is expected to continue throughout the next century (Figure 10.1).

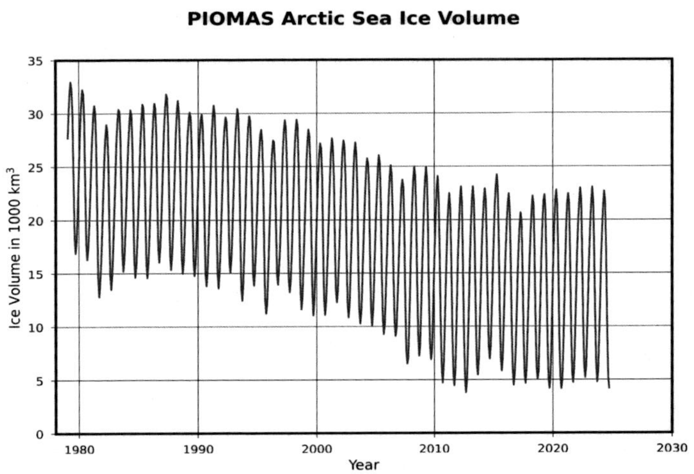

Figure 10.1. *Arctic experiences unprecedentedly low levels of snow and ice.*

Source: By Dominiklenne –Volume time series and uncertainties:Schweiger, A., R. Lindsay, J. Zhang, M. Steele, H. Stern, Uncertainty in modeled arctic sea ice volume, J. Geophys. Res., doi:10.1029/2011JC007084, 2011, Plot: own work, done with Python, Numpy, Matplotlib, based on published data of Polar Science Center, University of Washington,http://psc.apl.washington.edu/research/projects/arctic-sea-ice-volume-anomaly/data/, Public Domain, https://commons.wikimedia.org/w/index.php?curid=11321499.

10.4.1. Some Predictions on a Global Scale

Models predict that with every 2°F of warming, the Arctic sea ice cap is expected to decrease by 15% and the area covered by it at the end of summer (September) is expected to decrease by 25%. It is important to note that this reduction does not have any impact on the rise of sea levels. In the future, it is expected that the coastal Greenland and Antarctic ice sheets will continue to melt or slide into the water. The melting rate of the ice sheets in the 21st century could significantly contribute to the rise in global sea levels. There is an expectation that glaciers will continue to decrease in size. The anticipated acceleration of melting rates will worsen the impact of rising sea levels (Kattsov et al., 2005).

10.4.2. Important U.S Assessments

According to projections, there is expected to be a decrease of approximately 15% in the amount of snow cover across the Northern Hemisphere by the year 2100. Based on the models, it is predicted that snowfall will commence at a later time and melt at an earlier time. Consequently, the duration of the snow season is expected to continue decreasing. It is anticipated that several regions will experience a decrease in snowfall (Figure 10.2).

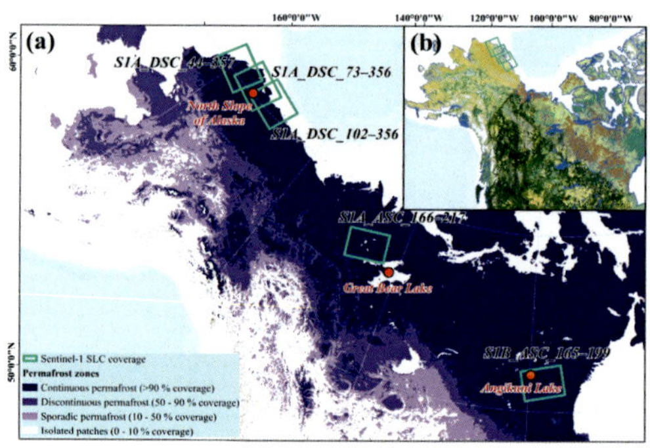

Figure 10.2. *Permafrost thawing in Alaska harms ecosystems, buildings, and infrastructure.*

Source: Guan S, Wang C, Tang Y, Zou L, Yu P, Li T, Zhang H. North American Circum-Arctic Permafrost Degradation Observation Using Sentinel-1 InSAR Data. Remote Sensing. *2024; 16(15):2809. https://doi.org/10.3390/rs16152809.*

10.5. ALTERATIONS TO FUTURE SEA LEVELS

Rising sea levels are caused by a combination of factors, including:

- The expansion of ocean water;
- The melting of mountain glaciers and ice caps; and

- The melting or flow of parts of the Greenland and Antarctic ice sheets into the ocean as a result of warming temperatures.

There has been a 7.5-inch rise in sea level worldwide since 1870. Sea level rise predictions for the next century are higher than those for the previous half-century, although these predictions vary by region. According to studies, there is a wide range of uncertainty when it comes to projecting the rise in global sea level by the year 2100. The estimated range is between 0.66 and 6.6 feet. In this particular instance, it is important to note that there is a need for a more instructional approach.

Research on the effects of climate change on Greenland's and Antarctica's ice sheets is ongoing, but studies on the role of thermal expansion, ice caps, and smaller glaciers in raising sea levels are more numerous and comprehensive. Current predictions indicate that 1.2–8 inches of sea level rise by the century's end will be caused by changes in ice sheets.

Future relative sea level rise for particular coastlines around the globe will be impacted by regional and local factors. Changes in land elevation due to processes like subsidence and uplift, for instance, determine relative sea level rise. If these geological processes from the past persist, a 2-foot increase in sea level by the year 2100 would lead to the following relative rise: 3.0 feet in New York City; 2.9% in Hampton Roads; VA, 3.5% in Galveston, TX; and 1.0% in Neah Bay, WA.

The relative increase of sea levels is also affected by regional shifts in wind speed and direction, water temperature and salinity, and the closeness to melting ice sheets (Kattsov et al., 2005).

10.6. ACIDIFICATION OF THE OCEANS IN THE FUTURE

Sea acidification in the future will have significant consequences for marine ecosystems. As carbon dioxide levels continue to rise, the pH of the oceans is decreasing, making them more acidic. This change in acidity can have detrimental effects on marine life, including coral reefs, shellfish, and other organisms that rely on calcium carbonate for their shells and skeletons. Additionally, acidification can disrupt the balance of marine food webs and impact the reproductive system.

According to predictions, there will be a decrease of over 30% in coral calcification rates if atmospheric CO_2 concentrations are doubled (Figure 10.3).

Figure 10.3. *Coral reefs may decline by 2050 due to ocean acidification and global warming from rising CO_2 levels.*

Source: López-Pérez A, Granja-Fernández R, Ramírez-Chávez E, Valencia-Méndez O, Rodríguez-Zaragoza FA, González-Mendoza T, Martínez-Castro A. Widespread Coral Bleaching and Mass Mortality of Reef-Building Corals in Southern Mexican Pacific Reefs Due to 2023 El Niño Warming. Oceans. *2024; 5(2):196-209. https://doi.org/10.3390/oceans5020012.*

BIBILIOGRAPHY

1. Ahmed, S. P. (2023). What is global warming and global warming potential? *Merchant Navy Decoded.* https://www.merchantnavydecoded.com/what-is-global-warming-and-global-warming-potential/ (accessed on 5 December 2024).
2. American Security Project-ASP. (2023). Climate security and national security, Climate change, ASP. Retrieved from: https://www.americansecurityproject.org/climate-security/ (accessed on 5 December 2024).
3. Arctic albedo–The science. (2023). *Beneath the Polar Sun.* https://beneaththepolarsun.org/sea-ice-albedo-feedback/ (accessed on 5 December 2024).
4. Bäckstrand, K., Kuyper, J. W., Linnér, B. O., & Lövbrand, E. (2017). Non-state actors in global climate governance: From Copenhagen to Paris and beyond. *Environmental Politics, 26*(4), 561–579.
5. Bakhtiari, F. (2018). International cooperative initiatives and the United Nations Framework Convention on Climate Change. *Climate Policy, 18*(5), 655–663.
6. Basics of climate change, US EPA. (2024). *US EPA.* https://www.epa.gov/climatechange-science/basics-climate-change (accessed on 5 December 2024).
7. Berhane, A. (2018). Climate change and variability impacts on agricultural productivity and food security. *Climate Weather Forecasting, 6*(240), 2.
8. Bernauer, T. (2013). Climate change politics. *Annual Review of Political Science, 16*, 421–448.

9. Boas, I., & Rothe, D. (2016). From conflict to resilience? Explaining recent changes in climate security discourse and practice. *Environmental Politics, 25*(4), 613–632.
10. Boege, V. (2018). Climate change and conflict in Oceania. *Toda Peace Institute, Policy Brief, 17*, 1–17.
11. Broome, J. (2008). The ethics of climate change. *Scientific American, 298*(6), 96–102.
12. Brown, O., Hammill, A., & McLeman, R. (2007). Climate change as the 'new' security threat: Implications for Africa. *International Affairs, 83*(6), 1141–1154.
13. Campbell, J. R. (2019). Climate change, migration, and land in Oceania. In *Climate Change and Conflict in the Pacific* (pp. 66–81). Routledge.
14. Climate Bonds Initiative. (2018). Climate adapted and resilient infrastructure–A green bridge to the SDGs. Retrieved from: https://www.climatebonds.net/2018/07/climate-adapted-and-resilient-infrastructure-%E2%80%93-green-bridge-sdgs (accessed on 5 December 2024).
15. Climate Change Science, US EPA. (2024). *US EPA*. https://www.epa.gov/climatechange-science (accessed on 5 December 2024).
16. Climate-ADAPT. (2023). Awareness raising campaigns for stakeholders' behavioural change–Discover the key services, thematic features and tools of Climate-ADAPT. Retrieved from: https://climate-adapt.eea.europa.eu/en/metadata/adaptation-options/awareness-campaigns-for-behavioural-change#:~:text=Awareness%20raising%20increases%20the%20knowledge,to%20respond%20to%20certain%20impacts (accessed on 5 December 2024).
17. Congressional Budget Office. (2022). Emissions of carbon dioxide in the transportation sector. Retrieved from: https://www.cbo.gov/publication/58861 (accessed on 5 December 2024).
18. Cubasch, U., Meehl, G. A., Boer, G. J., Stouffer, R. J., Dix, M., Noda, A., & Yap, K. S. (2001). Projections of future climate change. In *Climate Change 2001: The Scientific Basis. Contribution of WG1 to the Third Assessment Report of the IPCC (TAR)* (pp. 525–582). Cambridge University Press.
19. Dahal, K. R., Dahal, P., Adhikari, R. K., Naukkarinen, V., Panday, D., Bista, N., Helenius, J., & Marambe, B. (2022). Climate change impacts and adaptation in a hill farming system of the Himalayan region: Climatic trends, farmers' perceptions and practices. *Climate, 11*(1), 11. https://doi.org/10.3390/cli11010011.
20. Dahl, M. (2020). The feedbacks between climate change and air pollution–Exhaustion. *Exhaustion*. Retrieved from: https://www.exhaustion.eu/

resources/the-link-and-feedbacks-between-climate-change-and-air-pollution (accessed on 5 December 2024).

21. Daoudy, M., Sowers, J., & Weinthal, E. (2022). What is climate security? Framing risks around water, food, and migration in the Middle East and North Africa. *Wiley Interdisciplinary Reviews: Water, 9*(3), e1582.
22. Delbeke, J., Runge-Metzger, A., Slingenberg, Y., & Werksman, J. (2019). The Paris Agreement. In *Towards a Climate-Neutral Europe* (pp. 24–45). Routledge.
23. Dietz, S., Hepburn, C. J., & Stern, N. (2007). Economics, ethics, and climate change. *Ethics and Climate Change*, December 2007.
24. Division, E., & Nikolayeva, L. (2017). Climate change and security in Eastern Europe: Republic of Belarus, Republic of Moldova, Ukraine–Regional assessment. *United Nations Environment Programme*. https://wedocs.unep.org/handle/20.500.11822/23026 (accessed on 5 December 2024).
25. Durack, P. J., Gleckler, P. J., Landerer, F. W., & Taylor, K. E. (2014). Quantifying underestimates of long-term upper-ocean warming. *Nature Climate Change, 4*(11), 999–1005.
26. Energy Education. (2024). Polluter pays principle–Energy education. Retrieved from: https://energyeducation.ca/encyclopedia/Polluter_pays_principle (accessed on 5 December 2024).
27. Food System Primer. (n.d.). Food and climate change, Food system primer. Retrieved from: https://foodsystemprimer.org/production/food-and-climate-change (accessed on 5 December 2024).
28. Friman, M., & Strandberg, G. (2014). Historical responsibility for climate change: Science and the science–policy interface. *WIREs Climate Change, 5*(3), 297–316. https://doi.org/10.1002/wcc.270.
29. Garcia, D. (2008). The climate security divide: Bridging human and national security in Africa. *African Security Review, 17*(3), 2–17.
30. Gemenne, F. (2014). In quest to understand climate change and conflict, avoid simplification. *New Security Beat*. Retrieved from: https://www.newsecuritybeat.org/2014/03/quest-understand-climate-change-conflict-avoid-simplification/ (accessed on 5 December 2024).
31. Gemenne, F., Barnett, J., Adger, W. N., & Dabelko, G. D. (2014). Climate and security: Evidence, emerging risks, and a new agenda. *Climatic Change, 123*, 1–9.
32. Hickmann, T., Widerberg, O., Lederer, M., & Pattberg, P. (2021). The United Nations Framework Convention on Climate Change Secretariat as an orchestrator in global climate policymaking. *International Review of*

Administrative Sciences, 87(1), 21–38.

33. IGES. (2024). IGES and climate security: Asia-Pacific climate security project. Retrieved from: https://www.iges.or.jp/en/projects/climate-security (accessed on 5 December 2024).
34. Kattsov, V. M., Källén, E., Cattle, H. P., Christensen, J., Drange, H., Hanssen-Bauer, I., & Vavulin, S. (2005). Future climate change: Modeling and scenarios for the Arctic. In *Climate Change 2001: The Scientific Basis* (pp. 100–143). Cambridge University Press.
35. Knight, K. W. (2016). Public awareness and perception of climate change: A quantitative cross-national study. *Environmental Sociology, 2*(1), 101–113.
36. Knight, R. (2023). The science of climate change. *Citizens' Climate Lobby*. https://citizensclimatelobby.org/blog/policy/the-science-behind-climate-change/ (accessed on 5 December 2024).
37. Krampe, F., & Mobjörk, M. (2018). Responding to climate-related security risks: Reviewing regional organizations in Asia and Africa. *Current Climate Change Reports, 4*, 330–337.
38. Kuyper, J. W., Linnér, B. O., & Schroeder, H. (2018). Non-state actors in hybrid global climate governance: Justice, legitimacy, and effectiveness in a post-Paris era. *Wiley Interdisciplinary Reviews: Climate Change, 9*(1), e497.
39. Lazard, O., & Youngs, R. (2021). The EU and climate security: Toward ecological diplomacy. *Carnegie Europe, 12*.
40. Leary, N., Dokken, D., White, K., & Berger, A. (2001). Climate change 2001: Impacts, adaptation, and vulnerability. *Intergovernmental Panel on Climate Change*, 301–305.
41. Lindwall, T. (2024). What are the effects of climate change? *Natural Resources Defense Council*. Retrieved from: https://www.nrdc.org/stories/what-are-effects-climate-change (accessed on 5 December 2024).
42. Mäder, J. A., Staehelin, J., Peter, T., Brunner, D., Rieder, H. E., & Stahel, W. A. (2010). Evidence for the effectiveness of the Montreal Protocol to protect the ozone layer. *Atmospheric Chemistry and Physics, 10*(24), 12161–12171.
43. Martin, M. (2022). Climate action–United Nations sustainable development. *United Nations Sustainable Development*. Retrieved from: https://www.un.org/sustainabledevelopment/climate-action/#:~:text=Goal%2013%20calls%20for%20urgent,well%20below%202%20degrees%20Celsius (accessed on 5 December 2024).
44. Matthew, R. A., & Hammill, A. (2009). Sustainable development and

climate change. *International Affairs, 85*(6), 1117–1128.

45. Maystadt, J. F., & Ecker, O. (2014). Extreme weather and civil war: Does drought fuel conflict in Somalia through livestock price shocks? *American Journal of Agricultural Economics, 96*(4), 1157–1182.

46. McDonald, M. (2013). Discourses of climate security. *Political Geography, 33*, 42–51.

47. MEO. (2023). IOM chief raises concern over rise in Horn of Africa migration, MEO. Retrieved from: https://middle-east-online.com/en/iom-chief-raises-concern-over-rise-horn-africa-migration (accessed on 5 December 2024).

48. Munasinghe, M. (2011). Addressing sustainable development and climate change together using sustainomics. *Wiley Interdisciplinary Reviews: Climate Change, 2*(1), 7–18.

49. Nasiritousi, N., Hjerpe, M., & Linnér, B. O. (2016). The roles of non-state actors in climate change governance: Understanding agency through governance profiles. *International Environmental Agreements: Politics, Law and Economics, 16*, 109–126.

50. Noguer, M., van der Linden, P. J., Dai, X., Maskell, K., & Johnson, C. A. (2001). Climate change 2001: The scientific basis (Vol. 881, No. 9). J. T. Houghton, Y. D. J. G. Ding, & D. J. Griggs (Eds.). Cambridge University Press.

51. Pickering, J., McGee, J. S., Stephens, T., & Karlsson-Vinkhuyzen, S. I. (2018). The impact of the US retreat from the Paris Agreement: Kyoto revisited? *Climate Policy, 18*(7), 818–827.

52. Prinz, D. (2009). Contributor and victim–Indonesia's role in global climate change with special reference to Kalimantan. *Jurnal Sains Dan Teknologi Lingkungan/Jurnal Sains & Teknologi Lingkungan*. https://doi.org/10.20885/jstl.vol1.iss2.art5.

53. Prizzia, R., & Levy, J. (2018). Towards climate security and sustainable security in the Asia-Pacific region. In *Asia-Pacific Security Challenges: Managing Black Swans and Persistent Threats* (pp. 41–64). Springer.

54. Savaresi, A. (2016). The Paris Agreement: A new beginning? *Journal of Energy & Natural Resources Law, 34*(1), 16–26.

55. Singer, P. (2006). Ethics and climate change: A commentary on MacCracken, Toman and Gardiner. *Environmental Values, 15*(3), 415–422.

56. State of the Planet. (2023). Why climate change is a national security risk. *Columbia University's Earth Institute*. Retrieved from: https://news.climate.columbia.edu/2023/10/11/why-climate-change-is-a-national-

security-risk/ (accessed on 5 December 2024).

57. Stern, T. (2018). The Paris Agreement and its future. *The Brookings Institute, 5*(7).
58. StudySmarter UK. (2024). Climate change and sustainability: Relationship, StudySmarter. Retrieved from: https://www.studysmarter.co.uk/explanations/human-geography/economic-geography/climate-change-and-sustainability/#:~:text=Human%20development%20can%20negatively%20impact,issue%20that%20climate%20change%20presents (accessed on 5 December 2024).
59. UNDP Climate Promise. (2023). What is climate security and why is it important? Retrieved from: https://climatepromise.undp.org/news-and-stories/what-climate-security-and-why-it-important#:~:text=Climate%20security%20refers%20to%20the,of%20social%20tensions%20and%20instability (accessed on 5 December 2024).
60. UNDP. (2024). Sustainable Development Goals. Retrieved from: https://www.undp.org/sustainable-development-goals (accessed on 5 December 2024).
61. UNEP–UN Environment Programme. (2024). Climate change and security risks. Retrieved from: https://www.unep.org/topics/fresh-water/disasters-and-climate-change/climate-change-and-security-risks#:~:text=In%20recent%20years%2C%20the%20linkages,local%20levels%2C%20or%20even%20internationally (accessed on 5 December 2024).
62. UNICEF. (2022). Children are facing deadly drought in the Horn of Africa. Retrieved from: https://www.unicef.org/stories/climate-drought-horn-of-africa (accessed on 5 December 2024).
63. United Nations. (2007). The greatest threat to global security: Climate change is not merely an environmental problem. *United Nations*. Retrieved from: https://www.un.org/en/chronicle/article/greatest-threat-global-security-climate-change-not-merely-environmental-problem (accessed on 5 December 2024).
64. United States Department of State. (2023). Strengthening climate security through enhanced security sector governance. *United States Department of State*. Retrieved from: https://www.state.gov/strengthening-climate-security-through-enhanced-security-sector-governance/#:~:text=The%20United%20States%20recognizes%20that,the%20adverse%20impacts%20of%20climate (accessed on 5 December 2024).
65. US EPA. (2024). Causes of climate change. *US EPA*. Retrieved from: https://www.epa.gov/climatechange-science/causes-climate-change (accessed on 5 December 2024).

66. Velders, G. J., Andersen, S. O., Daniel, J. S., Fahey, D. W., & McFarland, M. (2007). The importance of the Montreal Protocol in protecting climate. *Proceedings of the National Academy of Sciences, 104*(12), 4814–4819.

67. Watanabe, M., Kamae, Y., Shiogama, H., DeAngelis, A. M., & Suzuki, K. (2018). Low clouds link equilibrium climate sensitivity to hydrological sensitivity. *Nature Climate Change*. https://doi.org/10.1038/s41558-018-0272-0.

68. Wfca_inanna. (2024). Wildfires and climate change. *WFCA*. Retrieved from: https://wfca.com/wildfire-articles/wildfires-and-climate-change/ (accessed on 5 December 2024).

69. WHO. (2024). Raising awareness on climate change and health. *World Health Organization*. Retrieved from: https://www.who.int/europe/activities/raising-awareness-on-climate-change-and-health (accessed on 5 December 2024).

70. Wilson Center. (2014). National security and climate change: What do we need to know? *Wilson Center*. Retrieved from: https://www.wilsoncenter.org/event/national-security-and-climate-change-what-do-we-need-to-know (accessed on 5 December 2024).

71. World Economic Forum. (2023). Climate change is becoming a security issue, says this expert. *World Economic Forum*. Retrieved from: https://www.weforum.org/agenda/2023/10/climate-change-national-security-defence/ (accessed on 5 December 2024).

72. Youngs, R. (2021). The EU's indirect and defensive approach to climate security–The EU and climate security: Toward ecological diplomacy. *Carnegie Europe*. https://carnegieeurope.eu/2021/07/12/eu-s-indirect-and-defensive-approach-to-climate-security-pub-84874.

73. Yusoff, K., & Gabrys, J. (2011). Climate change and the imagination. *Wiley Interdisciplinary Reviews: Climate Change, 2*(4), 516–534.

74. Zach Stein. (2024). Greenhouse gases, Definition & how to reduce them. *Carbon Collective Investing LLC*. Retrieved from: https://www.carboncollective.co/sustainable-investing/greenhouse-gases (accessed on 5 December 2024).

75. Zwolski, K., & Kaunert, C. (2011). The EU and climate security: A case of successful norm entrepreneurship? *European Security, 20*(1), 21–43.

INDEX

G

H

I